THESIS WITHOUT TEARS

A Guide for African University Students

Adonis & Abbey Publishers Ltd & Skylark Publications Ltd.

St James House
13 Kensington Square,
London, W8 5HD
United Kingdom

Website: http://www.adonis-abbey.com
E-mail Address: editor@adonis-abbey.com

Nigeria:
Suites C4 & C5 J-Plus Plaza
Asokoro, Abuja, Nigeria
Tel: +234 (0) 7058078841/08052035034

Year of Publication 2015

Copyright © John Kuada

British Library Cataloguing-in-Publication Data
A catalogue record for this book is available from the British Library

ISBN: 978-1-909112-54-4

THESIS WITHOUT TEARS

A Guide for African University Students

John Kuada

ADONIS & ABBEY
PUBLISHERS LTD

TABLE OF CONTENTS

CHAPTER ONE

BUILDING RESEARCH TRADITION IN AFRICAN UNIVERSITIES

Introduction

One of the objectives of thesis writing in many universities is to train students in doing academic research – i.e. learning to adopt a systematic approach to investigating issues in their chosen fields of study and solving problems that they have identified. These initial trainings prepare some of them for higher academic degrees – e.g. a PhD.

Many students in Sub-Sahara African (SSA) universities, however, find thesis writing a very frustrating undertaking. A few of them give up along the way due to the frustrations. This is understandable for several reasons. Writing an academic thesis for the first time is like learning a new language. It requires an adoption of a new mindset and the acquisition of new skills. My teaching experiences in Ghana and interactions with colleagues and students from different African and European universities have drawn my attention to a number of operational challenges that students face in their thesis writing process. Some of these challenges are more serious among African students due to the conditions under which their studies are undertaken.

First, some African students complain that their supervisors assume that they already have good research skills before coming to them. The supervisors therefore expect them to know what good research entails and complain about their students' inabilities to do "what is expected of them". The wide gap between supervisors' expectations and students' performance creates further frustrations in the thesis writing process, since interactions between students and supervisors are normally characterized by tensions and negative emotions.

Second, it has been observed that many African university teachers appear unwilling to accept methods that accommodate some of the difficulties that their students face in doing research. For example, many of them tend to see social science research as progressing in a linear fashion. They reject the iterative process of research that allows students to move back and forth between the various stages in the project work process, and they are not encouraged to seek alternative paths to solving their research problems.

Third, although many social science researchers are now cautious of the dangers of replicating theories and models developed within the Western context to other countries and communities, African students are not trained to show that type of awareness in their research. Students are not required to question the theoretical rationale and meta-theoretical assumptions underlying the research they conduct.

Fourth, those who have done empirical investigations in Africa are aware of the challenges that researchers face with data access in many research situations. There are serious problems regarding securing appropriate samples and ensuring that all data can be feasibly collected within limited research budgets and time. The problems of empirical data collection in many African countries are compounded by the relatively low response rates from certain segments of the population. For example, it has been noted that the higher the status and position of Africans the less willing they are to fill in questionnaires.

The challenges listed above may also reflect some fundamental weaknesses in the research infrastructure in African universities – i.e. the general shortage of research expertise and experienced supervisors, as well as inadequately equipped libraries with limited access to modern journals and the Internet (Szanton and Manyika, 2001; Biermann and Jordaan, 2007; Dowse and Howie 2013). The fact that the number of universities and institutions of higher education on the continent has increased significantly during the past two decades has further aggravated the situation.

Focus of this Book

These observations have motivated me to write this book. Nearly four decades of university-level teaching experience have convinced me that most university students start their thesis writing without prior research experience. They therefore need some initial hand-holding to take the first steps in the research competence development process.

I published a book on research methodology in 2012 and received useful feedback from colleagues and students. This book benefits from these feedbacks. I have included modified versions of some of the chapters that students have found exceptionally useful in this book.

My aim is to guide students through the entire thesis work process, and to offer them some clear, straightforward and practical advice that will help them avoid common pitfalls along the journey. I have African students in mind in the writing process. But I am certain that students in other parts of the world can also benefit from reading it. I have adopted a writing style that communicates directly to each individual student and therefore addresses the reader in a second person singular form throughout the book.

Structure

The book has ten chapters. Chapter one provides you with the background to the book. Chapters two, three and four seek to help you improve your understanding of the basics of scientific research, including the nature of science, criteria for judging your thesis and how to develop a good and feasible work process in order to fulfill your examiners' expectations.

Chapters five and six provide you with a general overview of the role of theories (and meta-theories) in your thesis. I argue in chapter five that everything in science is predicated upon some philosophical position, even though these issues may not be explicitly articulated in a research project. The chapter therefore draws your attention to ontological and epistemological issues in research (those of you who need deeper insight into these issues may consider reading my book on *paradigms and philosophy of science* written for doctoral students but which also discusses issues that Master's degree thesis writers may find useful.) Chapter six provides you with an overview of conventional classification of theories and how you can identify the most appropriate theories for your work. Chapters seven, eight and nine provide you with general guidelines on the choices you may have to make with regard to data collection methods and techniques. Chapter ten pulls the discussions together and provides you with a summary of the key points and guidelines.

CHAPTER TWO

CHARACTERISTICS OF A WELL-WRITTEN THESIS

Introduction

As a student, you must realize that your thesis is a piece of scientific work. As such, it must be accountable to "scientific" norms and standards that are widely accepted by leading scholars in your field of study. This means you must adopt systematic and methodical approach to your investigation. But scientific research also involves personal choices and these choices are, to some extent, influenced by individual researchers' personal convictions. These personal convictions will make some types of research questions, theories and methods more appealing to you than others. The purpose of this chapter is to guide you through these initial decisions. It first draws your attention to the expectations of your examiners and the criteria that they are likely to use to assess the quality of your work. It then discusses the contributions that your study may make to knowledge in your field of study.

What Examiners Are Looking For

Examiners of your thesis are required to assess the extent to which the thesis demonstrates your knowledge of the rules of scientific investigation – i.e. your ability to do an independent research in the discipline of your choice. The criteria discussed in the subsequent sections apply to any kind of academic thesis. But examiners apply them differently according to the academic level at which your thesis has been written – e.g. a bachelor's degree, a Master's degree, or a PhD level.

Most examiners agree that the research process leading to a good thesis must be guided by the following considerations:

1. Your project must be based on a definite research problem or issue – i.e. specific research questions. It is important to justify the value of your research by indicating that the problems that you seek to investigate has not already been answered sufficiently in the available literature in the field.

2. It must be informed by appropriate theories and concepts. All good thesis problems are based on established theoretical and/or conceptual frameworks.

3. Its scope and limitations must be clearly defined.

4. It must build on previous research, but must offer something new as well – i.e. contribute to knowledge in the chosen field.

5. It must be based on a robust methodology. That is, if it is an empirically based investigation it must have an appropriate approach to collecting and organizing data. It must also have an appropriate approach to analysing data.

6. It must offer an informed interpretation of results, and its findings must be consistent with the research question and implementation of the research design.

7. The thoughts presented must flow logically. That is, the underlying logic and supporting evidence presented in the thesis must be compelling.

8. The entire thesis must be seen as a professional piece of work. This means the text must be flawless, easy to follow and must give the reader a sense of completeness.

9. It may suggest directions for further research in the field. (See Knobel and Lankshear, 1999 for details)

Lovitts' Evaluation Criteria

Lovitts (2005) conducted a focus group study in the USA in 2003/2004 across 10 disciplines at 9 research universities. The results are presented in Table 2a and 2b. In the main, the conclusions from this study support the nine items listed above. They can also be seen as general criteria for evaluating any academic thesis in any part of the world. They describe characteristics that make a thesis either (1) outstanding, (2) very good, (3) acceptable, or (4) unacceptable. You can therefore apply them yourself to undertake a personal evaluation of your work.

Table 2.1a: The Characteristics of Outstanding and Very Good Thesis

Outstanding Thesis

- The research on which the thesis is based can be described as original and significant, ambitious, brilliant, clear, clever, coherent, compelling, creative, elegant, engaging, and exciting.

- The arguments in the thesis are focused, logical, insightful, and exhibit mature and independent thinking.

- The author asks new questions or addresses an important question or problem, and displays a deep understanding of a substantial amount of complicated literature.

- The study itself has a brilliant research design and uses or develops new tools, methods, and/or approaches to the investigation.

- The analysis is comprehensive, complete, sophisticated, and convincing; the results are significant and the conclusions tie the whole study together in a neat manner.

- The thesis as a whole pushes the discipline's boundaries and opens new areas for research.

Very Good Thesis

- The study can be described as solid, well written and well organized with strong, comprehensive, and coherent arguments.

- It has some original ideas, insight and uses appropriate (standard) theory.

- The research strategy demonstrates good normal science (i.e. well-executed research) and shows understanding and mastery of the subject matter

- The results are solid (but not extraordinary) and the author may miss opportunities to explore interesting issues and connections and therefore makes a modest contribution to the field.

Source: Lovitts, B., Academe, Nov/Dec 2005, p. 18-23

Table 2.1b: The Characteristics of Acceptable and Unacceptable Thesis

Acceptable Thesis

- The study discusses a question or problem that is not exciting; but the author demonstrates technical competence and shows the ability to do research.

- The author displays little creativity, imagination, or insight and has a weak structure and organization

- The author reviews the literature adequately (i.e. demonstrates adequate knowledge of the literature) but is not critical of it or does not discuss what is important.

- The author uses standard methods, performs unsophisticated analysis and produces predictable results that make modest contribution to the field.

Unacceptable Thesis

- The thesis is poorly written and contains spelling and grammatical errors.

- The author plagiarizes and/or deliberately misreads or misuses sources.

- The author does not demonstrate understanding of basic concepts, processes, or conventions of the discipline and looks at a question or problem that is trivial, weak, or unoriginal.

- The arguments in the thesis are weak, inconsistent, self-contradictory, unconvincing, or invalid.

- The data used in the study are flawed, wrong, false, misinterpreted, unsupported, or are characterized by exaggerated interpretations.

- The entire work reflects limited professionalism.

Source: Lovitts, B., *Academe,* Nov/Dec 2005, p. 18-23

Contribution to Knowledge

The evaluation criteria discussed above emphasize the need for a good thesis to move knowledge in your chosen field of investigation forward to some extent. Most thesis guidelines will tell you that your thesis must strive to make three types of contribution to knowledge:

1. Contextual/theoretical contribution
2. Empirical contribution
3. New theory development

The conceptual or theoretical contributions may be of the following types:

1. Conceptual combination
2. Conceptual expansion and reframing

Conceptual combination is the idea of linking two or more existing concepts - e.g. social responsibility + corporations have been combined to produce *corporate social responsibility*; or a combination of entrepreneurship + societal issues has resulted in the creation of the concept of *social entrepreneurship*. The ambition here is to create new concepts that can highlight interesting new ways of understanding existing social phenomena. Doing so requires imaginative and reflective thinking, and this can start a new strand of research.

Conceptual expansion/reframing extend a concept beyond its core use to match a new situation. This gives new or better informative description to an existing phenomenon. An example is "greenhouse effect", a concept that uses the greenhouse analogy to explain the environmental challenges of the 21st century.

You may need to read a wide variety of previous studies from different disciplines to be able to offer novel conceptualizations that provide richer insights into your subject. For this reason, many students are reluctant to make conceptual and theoretical contributions the overall goals of their study.

As we shall argue in subsequent chapters, researchers often develop abstractions from the messy detail of the social world they observe or investigation by the use of theories. These theories serve one or a combination of the following three purposes in research:

1. They provide frameworks for analysis;
2. They provide efficient methods for developing specific fields of research; and
3. They provide insights into frequently messy social phenomena.

As a researcher you may seek to build a theory through four steps: reflections on previous theories, observation of real life situations (phenomena), categorization of these situations, and identifying associations between variables/elements that describe the situation. That is, first you may reflect on what previous scholars have observed and noted about the phenomena; then you direct your attention to observing the focal phenomena, documenting what you have observed and measuring them (if necessary or possible). This process allows subsequent researchers to verify the authenticity of your theory building process. In the next stage the phenomena are classified into categories in terms of their attributes. This part

15

of the process is usually described as the creation of typologies or frameworks. In the fourth step, you explore the association between the category-defining attributes and the outcomes observed. If you do these in an elegant way, you will be in position to provide additional theoretical insights into the phenomena.

It is, however, important to bear in mind that all theories are, to some extent, incomplete and provisional - i.e. they are, of necessity simplifications of reality. That is, our understandings of how the world works at any given point in time in any given situation are never foolproof. But although these simplifications necessarily sacrifice some accuracy, they are valued for their elegance, insight, and usefulness in moving our understandings forward.

You may also consider a relevant contribution of your study to be one of examining the appropriateness of models, concepts and theories developed and applied in Western contexts to the operational contexts of your country. This type of project will require an empirical analysis within your chosen field of study, based on some specific concepts, models and theories.

In addition to testing the appropriateness of the concept, models and theories within a different context, you may also add to the empirical knowledge in the field by providing additional confirmation to the theories. It is, however, possible for these kinds of studies to produce dissenting evidence (i.e. not confirming the theories) and thereby challenge the validity of the existing knowledge.

In sum, right from the beginning of your study, you must have an idea about the quality of work you intend to produce and what kind of contribution you intend to make to your field of investigation. This will help you gauge the amount of effort

16

and resources you will need for your study. It is also useful to use the set of criteria listed in this chapter to do a personal assessment of the quality of your work in order not feel disappointed with your grades.

CHAPTER THREE

THE THESIS WORK PROCESS

Introduction

The discussions in chapter two suggest that a very important first step in the thesis work process is choosing the right topic and research issues that you want to investigate and determining the amount of effort your investigation entails. This chapter seeks to help you make these important decisions. It first discusses the iterative nature of a thesis work process. I know from experience that as you read further about the issues you intend to investigate and continue work on the problem you have initially formulated, you will discover newer dimensions and perspectives to the problem. This will make it necessary for you to modify the problem. It then discusses decisions regarding your choice of research themes and topics and to guide you through your problem formulation process. It also guides you on what to do when you get stuck in the process due to unforeseen circumstances.

Research as an Iterative Process

A research process is not a straight line. It is iterative. You must therefore develop your own logic of investigation. This will help you to determine issues that are essential and guide you to the sources of information required for your analysis. As your thesis progresses, you will acquire new knowledge and improve your perception of the issues that you investigate. You will then be able to take a fresh look at some of the ideas and viewpoints that you have earlier expressed and modify them in the light of your improved knowledge. Even your research questions (i.e. problem formulation) are likely to change as new information is acquired about the issues that you are investigating.

Start writing drafts of the various chapters of your thesis as soon as you have agreed on the research issues to investigate. This will give you the opportunity to think through your arguments several times during the thesis work process. If you begin the writing process only after all data have been collected you will realize that you will run out of time and you cannot read through the thesis carefully before submitting it. This can have disastrous consequences for the quality of the thesis.

The iterative process is illustrated in Figure 3.1. It shows that the initial steps in a research process is to locate gaps in existing knowledge and understanding, and frame these gaps clearly and concisely as problems and questions. The process continues with designing systematic, methodical, and reasonable ways of exploring the identified gaps, and building on existing concepts, theories, and methods to move knowledge forward. This process is described in details in the sections below.

Figure 3.1: A Schematic Illustration of a Thesis Work Process

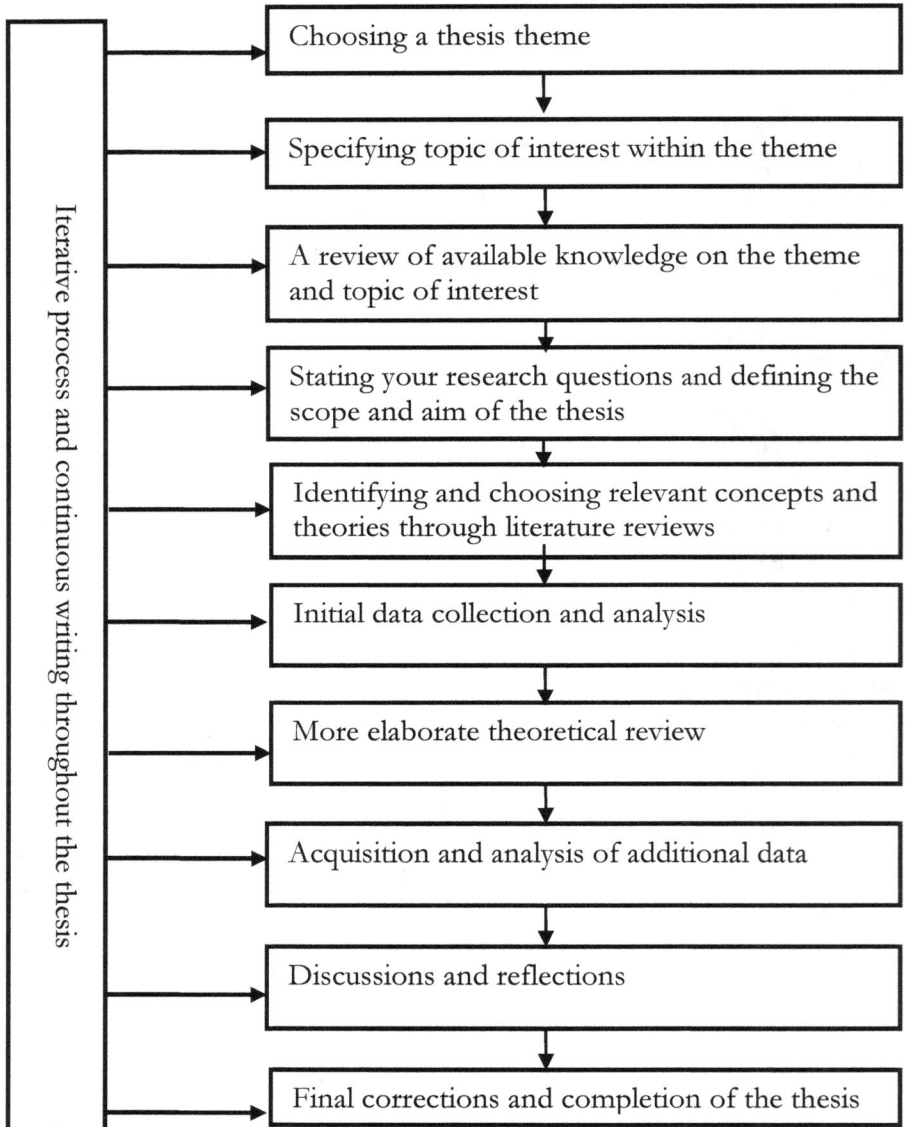

Iterative process and continuous writing throughout the thesis

- Choosing a thesis theme
- Specifying topic of interest within the theme
- A review of available knowledge on the theme and topic of interest
- Stating your research questions and defining the scope and aim of the thesis
- Identifying and choosing relevant concepts and theories through literature reviews
- Initial data collection and analysis
- More elaborate theoretical review
- Acquisition and analysis of additional data
- Discussions and reflections
- Final corrections and completion of the thesis

Source: Kuada, 2012 (p.36)

Choosing the Topic

Having decided on the theme of investigation you can proceed to choose a topic that captures your interest. The topic must be considered relevant and intellectually stimulating. That is, you must have a strong desire to work on that topic. Apart from your personal desire to work on the topic, the problem must be significant in some way – i.e. it must not be trivial or already solved.

Fisher (2010) provides a six-stage process model to guide students in choosing a thesis topic. These are reproduced in box 3.1 for a quick reference.

Box 3.1
Process for Choosing Research Topic

1. Identify a broad topic of interest
2. Determine the scope
3. Discuss the key issues, puzzles and questions that arise from the topic
4. Map and structure these issues to see their interconnections
5. Reflect on the relevance/research potential of each of the issues
6. Frame the research questions you would like to address

Source Adapted from Fisher, Colin (2010) *Researching and Writing a Dissertation* (Essex, FT Prentice Hall page 35)

After deciding on the topic, you must also discuss the scope of your research – i.e. what to include and what to leave out of the thesis. Like other parts of the process, this is also a tentative decision. However, doing so entails brainstorming and using lecture notes and other readings and combing them with your personal observations and reflections.

Research Statements and Questions

Having agreed on the topic, the next stage in the thesis work process is to define the research questions (i.e. possible problems that will be worth investigating and/or solving within the chosen topic). Ideally, research problems must be defined in specific terms in order to avoid creating doubts in the minds of the group members as to the focus and scope of the thesis. But such a clear-cut definition may be nearly impossible in social science research. Again, the problem definition must be considered as an iterative process, where new knowledge about the problem improves your perception and the clarity with which the problem can be defined.

You may initially arrive at a "working definition" of your problem - i.e. a tentative problem description that highlights the central issues of investigation. For example, if you are a business student who is interested in the analysis of pharmaceutical industries, you may want to examine *the degree of concentration* within that industry and the factors that influence relationships of companies within the industry. Or if you are interested in export marketing, and may want to investigate the market selection process of small companies in your country or of companies within a specific industry.

The following questions may provide you with some inspiration in your problem formulation process:

1. What is the background to the problem - i.e. how did it come about?
2. What has been the focus of past studies done in this area?
3. What have these earlier studies ignored or paid limited attention to?

22

4. What knowledge gap can I identify from reading these past studies?
5. What are the main research issues I want to focus my attention on?
6. What makes these issues interesting and relevant?
7. What do I hope to accomplish in the thesis?
8. What are my strengths (and limitations/constraints) in conducting this investigation?
9. Can these issues be investigated within the specified time limits?
10. Can I have access to relevant data?

Discussions along these lines will help you provide a strong justification for your thesis – i.e. demonstrate its relevance to some specific stakeholders in an organization or society or justify the enhancement of knowledge that the study seeks to provide. This will also help you avoid choosing a problem area that may be interesting but too broad to be handled in a satisfactory manner within the short span of thesis time. Bear in mind that by focusing attention on selected issues you automatically ignore other issues. This is normal. It is, however, a good idea for you to inform your readers that you are aware of this limitation.

Building on Previous Knowledge

I have repeatedly emphasized the importance of reading previous research in order to establish the legitimacy of your own study. I would like to discuss this issue in greater details in this section in order to underscore its importance and to provide you with some guidelines on how to proceed. The question you need to ask yourself again and again in the process is this:

Does your thesis add value to current thinking in your chosen field of investigation? How significant is the added value? Another question they are likely to ask is this: "what is new in this study?" That is, what contribution are you making to current knowledge? The expectation for a Master's degree thesis is usually modest. You are not expected to produce anything revolutionary. But you must be able to offer a convincing solution to what you find to be lacking in your chosen field of investigation.

The literature review should be more than a catalogue of the literature. It should contain a critical, analytic approach, with an understanding of sources of error and differences of opinion. The literature review should not be over-inclusive. It should not cover non-essential literature or contain irrelevant digressions.

Whetten (1989) observes that research must be both comprehensive and parsimonious. Comprehensiveness here means that you must endeavour to include all relevant factors in your research in order to understand the phenomenon that you intend to investigate. Without comprehensiveness, you may not be adding anything new to the field. Parsimony, on the other hand, requires you to include only those factors that make substantial contribution to the field. The requirements of comprehensiveness and parsimony therefore constitute a dilemma that you must try to manage. This means you must read extensively on the theme of your research in order to have good knowledge of what other people have written and how to build on the existing knowledge.

Again, remember that you must position your research in relation to the existing body of literature – i.e. build on previous research done in the subject. This will help you identify issues that have not been covered by previous researchers and therefore provide a justification of your research. A strong justification for the research motivates readers to read beyond the opening paragraph of the thesis. This is one of the reasons why you must read a bit about the topic, noting the mainstream thoughts and issues discussed as well as the different viewpoints expressed by other scholars on the issue. It will also be useful for you to involve your supervisor in the discussions and to draw on other academic staff members that may have some knowledge of the issues that you consider worth investigating.

Having agreed on the issues to investigate you can now plan the subsequent stages of the research, including the theoretical foundations on which to base your investigations, methods to adopt to fulfil the objectives, and how to structure the entire thesis. I will discuss these in details in part two of the book.

Research Strategy and Resources

Considering the fact that students normally have limited time and resources for their thesis, you may need to discuss these resource limitations while choosing a problem for investigation. For example, there are some issues in Economics that are highly interesting, important and may be relevant for investigation within the current economic situations but too resource demanding for you to undertake. From a learning perspective, it will be pedagogically more rewarding for you to work on a problem that is less ambitious in scope and novelty than to set out on an ambitious investigation that you may abandon half-way. You must therefore consider your resource limitations when defining the problem in order to avoid unpleasant surprises.

The above statements suggest that preliminary discussions of appropriate methods of investigation must be undertaken already when discussing the focus of the thesis. It is important for you to get some idea about which kinds of data are required for a satisfactory work on the various problems of interest, the sources of such data, methods of data collection, and anticipated problems in collecting them. These considerations should influence your problem formulation.

In some cases the data collection process may prove more difficult than you anticipate even after previous elaborate discussion. You must not discontinue the thesis on that account. Discuss the difficulties with your supervisor and work out an acceptable approach that will enable you to finish the thesis.

Objectives of Your Study

It is important for you to state the aim of your thesis very clearly. This specifies what your readers should expect from reading the thesis. You can choose between a *descriptive* and a *normative* type of research. As the name implies, the aim of a descriptive research is to provide a description of a particular problem under investigation. That is, the thesis provides a clear picture of the issues investigated. For example, your thesis may be about European companies' attitude to investment in developing countries. Such a thesis may provide information on the number of European companies investing in African countries, the distribution of their investment in terms of size, geographical location, sector, and product. You may also describe the investment decision making process as

26

well as the underlying reasons for making such investments. Further investigations may also examine whether there is any relationship between size of company, industry of operation, and investment decision, etc. Such an investigation can form the basis for forecasting European foreign investment in African countries in the subsequent years. This kind of thesis has a descriptive objective (or ambition), because it describes what is happening, how it is happening and what is expected to happen in the future, based on what we know today.

On the other hand, a *normative* research thesis provides guidelines for decision making. That is, the thesis outlines what a rational decision-maker should do under the identified conditions in order to attain a given objective. For instance, if the thesis which examines investment in African countries had a normative aim, you will identify mistakes made by European companies in their investment decision making process and discuss the reasons for committing such errors. The thesis would also present suggestions for solving the problems in order to help the European companies make optimal use of their investment resources.

A normative study will have a descriptive part, which forms the foundation for the strategy or actions proposed by the group. This means that you will use its analysis of the situation to justify the guidelines you recommend. Many supervisors of business-related student thesis encourage their students to have both descriptive and normative ambitions for their theses.

The Role of a Supervisor

It has been hinted above that working on a thesis is an independent student exercise under the supervision of an academic staff member. The role of the supervisor is one of a facilitator rather than a director. He guides the learning process by asking critical questions and making suggestions that help you to re-examine your thoughts on the problem of investigation. It is not his responsibility to define the problem of investigation, find the relevant literature or edit the work. If he does any of these things, you must see it as a kind gesture rather than an obligation. The supervisor's contribution to your learning process will be greatly facilitated if you present your views clearly and specify the problems and/or doubts you have. By involving the supervisor in the discussion, you may be able to improve your perception of the issues and move forward with your work.

Most supervisors would like you to present them with a summary of your discussions and the issues on which you require their opinion. This may cover a few pages. You must avoid sending 20-30 pages to your supervisor each time you request for a meeting without clearly specifying which issues/problems you would like the supervisor to discuss with you.

Good supervisors provide students with feedback that helps them think through their arguments, strengthen their logical foundations, ensure greater clarity in their writings, especially with respect to the language and concepts used in their chosen field or discipline. Good feedback will build your confidence at later stages in the thesis writing process and encourage you to take more critical stance to your work.

CHAPTER FOUR

STRUCTURE OF YOUR THESIS

Introduction

One of the main criteria for evaluating your final report is the relevance of the materials that you have presented in each chapter and the logical flow of your arguments. Experience shows that some students place greater emphasis on the size of the report (i.e. the number of pages) than on the relevance of the contents to the problem that they set out to investigate. Very often, the descriptive chapters that are aimed at providing background information swell up during the writing process and assume prominence over the analytical chapters. As a result, although the final report covers several pages, the substantive discussions become thin and superficial. It is therefore important for you to pay attention to each section and chapter of the thesis and make sure that the purpose is clearly communicated to the reader.

This chapter provides you with some guidelines in the design of the entire thesis. It draws your attention to the role that each chapter can play in the thesis. Some universities and departments provide their students with clear guidelines on how to structure student thesis. In the absence of such guidelines, the structure presented in this chapter should offer you a useful guide.

A General Thesis Structure

Box 4.1 provides you with a list of items included in a thesis. This constitutes a generic thesis structure. Different universities and departments may deviate from this general structure by specifying a sequence of items that serves the purpose of the programme that they offer. Universities also differ in terms of text layout, section headings and sub-sections that they recommend their students to use. Follow the guidelines offered at your university.

Box 4.1
A General Thesis Structure

- Title page
- Table of contents
- Abstract or executive summary
- Acknowledgement
- List of tables (if any)
- List of acronyms (if any)
- Introduction
- Main body of the thesis
  This will usually contain several chapters including methodological, theoretical, and empirical chapters as well as discussion and reflection chapters
- Summary and conclusions
- References
- Appendices

Title page

The title page provides the following information

1. The title of the thesis
2. The name of the student

3. The name of the study programme and the semester
4. The name of the department, and university

Table of Contents

The table of contents provides a list of the chapters and main sections in the thesis as well as the pages on which they appear. Some universities specify different levels of headers that students must use in their thesis. In the absence of specific guidelines, students may use three levels of headers. The first is the title of the chapter. The second is the title of each section (also called *A-heads)*, and the third is the title of each sub-section (also called *B-heads*). The contents normally contain only first level headings, and in some cases second level headings.

Abstract/Executive Summary

Abstracts and executive summaries are written after the report has been completed. An abstract covers just half a page. It provides a summary of the whole thesis, highlighting the reasons for the thesis, the research design, the main findings and the conclusion. An executive summary serves the same objective but is a bit longer (2 pages maximum) and is usually written by business students who would like to provide executives with a summary of their investigations. These executive summaries place emphasis on the main findings, conclusions and recommendations.

The choice between abstract and executive summary is usually determined by the preferences of the programme director and/or supervisor of the thesis. You are therefore advised to read the guidelines for your programme to determine whether abstracts or executive summaries are required.

Acknowledgements

The acknowledgement section offers you an opportunity to express your gratitude to those who have supported you in the research process. These may include organisations, companies, managers and other officers that have granted you interviews or supported you by providing other forms of data. You may also offer thanks to persons and organisations that have granted you financial assistance in connection with the thesis.

List of Tables

As an aid to the reader, it may be a good idea to provide a list of all Tables included in the thesis for a quick overview. But this is not obligatory.

List of Acronyms

Acronyms are made up from the first letters of official names or titles. Examples include institutions such as the UN, NATO, WTO, EU, and OECD. You may also create your own acronyms as short versions of names of organisations that appear in the thesis. It is purposeful to provide a list of these acronyms and what they officially cover so that the reader can make quick references to them when he/she is in doubt.

Introduction

The introduction of the thesis provides the background for the research you seek to undertake. It tells the reader what is known about the subject, and what is not known, and explains how the research questions can add knowledge and understanding to the topic. When writing your introduction, you must specify the domain of your research in the opening paragraph of the introduction. This paragraph should highlight the importance of the research topic and why it deserves academic attention. The immediate justification of the research topic will motivate readers to read beyond the opening paragraph.

The second paragraph should further help you to develop the research problem. As noted in chapter 3, researchers normally do so by locating the issues of investigation in the existing literature on the topic – i.e. by providing a brief but focused review of the available literature. This paragraph indicates the current state of knowledge in the area and what is important to know but is not yet known. Remember that a study cannot be justified on the grounds that it has not been done previously.

You will normally start writing the introductory chapter of the thesis right from the beginning of the thesis work. But remember that you cannot fully introduce the thesis until you have finished your work. The initial drafts of the introductory chapter will provide you with a direction or roadmap for the thesis. When the thesis is completed, read through these drafts once again and revise them to reflect what you have actually done before submitting your thesis for examination. That is, you must be sure that what is in the introduction is consistent with the various elements in the entire thesis.

Body of the Report

The main body of the thesis will typically contain a number of chapters that discuss the following issues:

1. The *Methodology Chapter.* You will present your research design in this chapter. The research design explains the master plan of the research – i.e. how to conduct the research and the methods used. That is, you must specify which kind of data you will collect, why, where and how you will collect them, and how you will analyse the data in order to answer the research questions. (See chapter 5 of the book for a more detailed discussion. Additional information can be obtained in chapters 8 and 9 regarding choice of methods).

2. *The Theoretical Chapters:* These are chapters that highlight your understanding of the existing body of theories on which your study is based. This may contain several chapters, depending on the nature of the thesis. (See chapter 6 of the book for a more detailed discussion).

3. *Empirical Chapters:* These are chapters that present an analysis of the data as well as the major findings from the study.

4. *Discussions and Reflections:* You should devote a chapter to the discussions of the implications of your findings as well as some reflections on the thesis process itself. This chapter allows you to take a final look at what you have done. First, you will discuss the implications of your findings for the various stakeholders that the thesis concerns – governments, organizations, companies and/or various groups in the society. Second, you will also reflect on how you have conducted the investigations, the difficulties that you have faced, and

how you have addressed these difficulties. Third, you may specifically reflect on your choices of theories and method, noting the consistencies between them.

Summary and Conclusions

Some students have a tendency to write their conclusions hurriedly as if they do not expect them to be read with any seriousness. You must pay attention to your conclusions. Most examiners may take a look at the introductory chapter of the thesis and then hold the contents against the conclusions in order to ascertain whether you have delivered what you promised in the introduction. That is, examiners read conclusions with substantial attention.

You may start your conclusion chapter by briefly summarising the thesis, highlighting the main assumptions and findings. There must be consistency between what you have written in the introduction and the conclusions. That is, the conclusion must reflect the extent to which you have been successful in fulfilling the objectives set out for the thesis. Box 4.2 provides you with some guidelines on how to write conclusions.

Box 4.2
Guidelines for Writing Conclusion

- Are the conclusions related to the focus of the investigation (i.e. the research problem)?
- Have we solid arguments and evidence to bear the conclusions?
- What are the limitations in our analyses and how do they influence our conclusions?

References

References are very important. They establish the credibility of your arguments and indicate their main sources. Social science researchers have developed standard procedures for writing references. If your references deviate very much from the standard guidelines, this will reduce the quality of the thesis. This section of the chapter provides you with some guidelines on how to write references.

References can be classified into two types. There are citations in the text (or footnotes) and there is a list of references or bibliography at the end of the thesis. Most universities recommend that students use the Harvard style citation. I have provided some examples of the Harvard style referencing below as a guide.

1. If the article or book you are referencing is written by one author you must list the author's name and the year of publication as in the example below.

> Kuada (1994) suggests that culture impacts managerial

2. If you are quoting the exact words from the author you must add the page on which the quote appears as shown below:

> Kuada (1994:50) states that "an individual manager's behaviour will be determined by influences from his socio-cultural and organizational environment".

3. If the reference is from two or three authors you must write the last names of all the authors followed by the year of publication as shown in the example below:

> Kuada and Sørensen (2000) suggest that internationalization of firms can be viewed from both upstream and downstream

4. If the reference is from three or more authors, you write the last name of the first author followed by *et al* (in italics) and the year of publication. But you must remember to write the names of **all** the authors when preparing the list of references at the end of the report. See the example in the box below.

> Blumberg *et al.* (2005) provide a detailed discussion of mixed methods research.

5. You must present a list of all the references you have used in your thesis at the end of the report. The list must be presented in an alphabetical order using the last names of the authors. The bibliography at the end of this book illustrates how to write your list of references.

You can read more about the Harvard style referencing from this link:

http://iskillzone.uwe.ac.uk/RenderPages/RenderCons
tellation.aspx?Context=10&Area=8&Room=25&Cons
tellation=40

Citing Sources from the Internet

There are no major differences between citing an Internet source and citing a book or a journal article. But you must remember to include the URL (the website address) at the end of the citation followed by the date of access – i.e the date you accessed the Internet source. It is advisable to copy and paste the URL to be sure that the address you have given is accurate. Most supervisors are uncomfortable with citations of documents for which no authors have been named. Please check with your supervisor, if you intend to use such sources and receive a clearance from him/her before you do so. Furthermore, some supervisors do not consider Wikipedia a reliable source that students can use. Again, check this with your supervisor.

Deciding on the Contents of your Chapters and Paragraphs

The list of questions presented in Box 4.3 can provide you with some guidelines on how to structure the chapters and paragraphs in your thesis.

Box 4.3
Guidelines for Structuring Chapters and Paragraphs

- What kinds of information should I present in the chapter I am now writing?
- What is the relationship between this chapter and other chapters in the report?
- Have I drawn the reader's attention to the content and structure of the chapter?
- Can I justify the inclusion of the ideas and materials I am presenting - i.e. how detailed should the presentation be?
- Has the central point come out clearly in my writings?
- Have the discussion in the chapter covered the mainstream viewpoints on the issue, or have I been too narrow in my selection of viewpoints? That is, have I adequately discussed the competing viewpoints?
- Am I critical enough in this presentation, or am I merely reproducing other peoples' viewpoints – i.e. without a critical assessment of their relevance to the issues I am investigating?
- Have I made my personal standpoints on these issues clear enough in the paragraphs, sections and the whole chapter?

CHAPTER FIVE

RESEARCH DESIGN

Introduction

Research design is usually described as *the master plan* or *blueprint* of a piece of research. It provides you with both the framework and road map for the research. That is, a good research design should provide a logical sequence of activities that allows the reader of your thesis to see the connections between the research questions that you have posed in the introduction chapter of the thesis, the approach that you adopt to address the questions, the assumptions underlying your approach, how you collect and analyse your data, as well as your findings and conclusions. Very often there will be more than one kind of theoretical approach, and more than one conceptual framework, that you may consider using to produce a coherent research design. These choices you make require careful deliberations.

The meta-theoretical (or paradigmatic) foundation of the thesis has a strong influence on the overall strategy of the thesis. But you are allowed to use whatever strategies, methods or empirical materials at hand in order to fulfil the objectives of your investigations. That is, you can piece together combinations of methods that serve your research purpose. The important thing for you to do is to explain what you have done and why you have made the various choices in your research design. You must also be aware of the limitations that the various choices impose on your research, and discuss these limitations in your thesis.

This chapter draws attention to the fundamental concepts used in the literature to describe the issues that you should consider with regard to research design. It also provides you with guidelines on how to decide on an appropriate research approach.

The Four Levels of Understanding

Most research methodology textbooks in social science identify four levels in a research design process. These levels feed into each other as illustrated in Figure 5.1.

```
┌─────────────────────────┐
│ Philosophical/Theor     │
│ etical Viewpoints       │
│                         │
│ Discussing issues of    │
│ ontology                │
└─────────────────────────┘
            ↘
        ┌─────────────────────────┐
        │ Epistemological Choice  │
        │                         │
        │ Views on how knowledge  │
        │ about the research should│
        │ be understood           │
        └─────────────────────────┘
                    ↘
            ┌─────────────────────────┐
            │ Methodological Decisions│
            │                         │
            │ Discussion of overall   │
            │ approach to the research│
            └─────────────────────────┘
                        ↘
                ┌─────────────────────────┐
                │ Choice of Methods and   │
                │ Techniques              │
                │                         │
                │ Description of data     │
                │ collection tools and reasons│
                │ for their choice        │
                └─────────────────────────┘
```

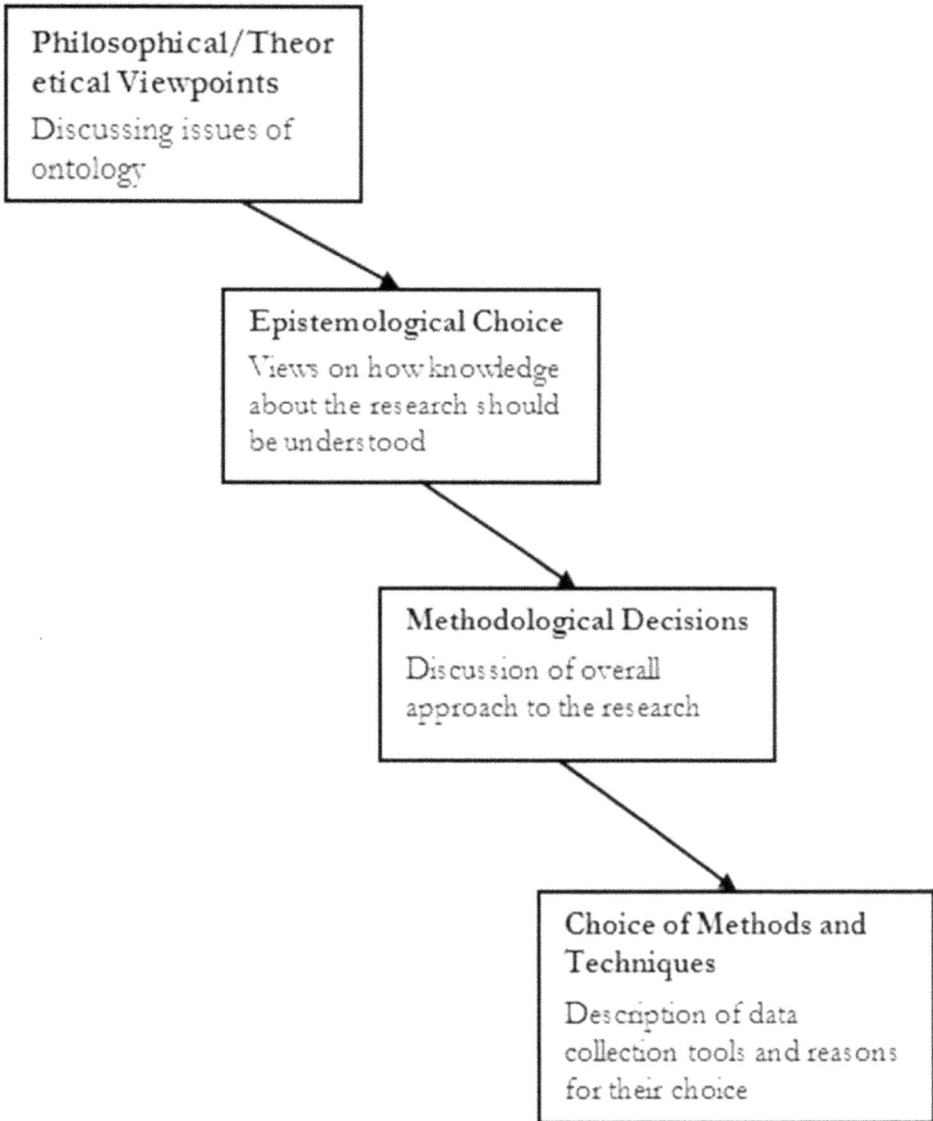

Figure 5.1: Structure and Levels of Discussion in a Methodology Chapter

41

Level 1: The Philosophical and Theoretical Level

Ontology is a term used by philosophy of science scholars to describe the nature of what the researcher seeks to know something about – i.e. the "knowable" or "reality". The social world that social science scholars investigate is usually seen from two broad perspectives. For some scholars the social world is real and external to an individual human being and therefore imposes itself on his consciousness. Other scholars hold the view that every individual creates his own social world. To them the social world is subjectively constructed and therefore a product of human cognition.

Ontology also relates to how researchers see the relationship between human beings and their environment – i.e. researchers' view of *human nature*. Some researchers see the social environment as being outside the individual. Other researchers hold the view that human beings and the social environment co-determine each other.

Thus, assumptions that you make about human beings and the environment will define your perception of reality. This perception, in turn, underlies what you will consider as a "truth" and how knowledge about this "truth" should be acquired.

Level 2: Epistemological Level

Epistemology is a term that describes the nature of knowledge and the means of knowing – i.e. *"how we know what we know"* or what we conceive as a "truth". Some scholars hold the view that it is possible for them (as external observers) to "know" the truth about a specific social world. Others maintain that the social world can only be understood by occupying the frame of reference of the individual actor whom the researcher seeks to study. That is, the social world must be studied "inter-subjectively".

Level 3: Methodological Approach

Methodology describes the reasons underlying the choice and use of specific methods in the research process, - i. e. how you may go about gaining the knowledge you desire. It involves understanding the assumptions underlying various research methods and the criteria to use in deciding which of them to use. That is, they provide you with theories that enable you to justify the methods that you choose to solve your problem.

For example, if you assume that the social world can be objectively observed from outside, you will adopt a methodology that focuses on an examination of relationships as universal laws. But if you assume that the social world can only be understood by obtaining first-hand knowledge of the persons under investigation, you will opt for a methodology that focuses on individuals' interpretations of the world as they experience it.

Level 4: Methods and Techniques

The term *"research methods"* is used to describe all those methods and techniques that you may consider to use during the course of studying your research problem. In other words, they constitute the tools for doing a piece of research. Thus, when writing the methods section of your methodology chapter you must describe the specific data collection methods and techniques you may adopt in your study. You must also inform your readers about the problems you faced in the data collection process and how you solved those problems. For example, if you choose to conduct interviews, you must indicate what interviewing techniques you have used and in what sort of setting the interviews have been conducted. You must also show the link between the methods selected and the problem formulation as well as the consistency between the methods and the three levels of understanding described above – i.e. the ontological perspectives, the epistemological considerations and the methodological approaches (see Crotty, 1998 for elaboration).

Exclusive versus Complementary Approaches

There is a controversy in the research methodology literature as to whether researchers can see "reality" only from an objective (i.e. external) or a subjective (i.e. socially constructed) perspective or whether reality can be seen from both perspectives in the same thesis. In other words, should researchers consider the two views of reality mutually exclusive (i.e. they cannot be combined) or should researchers see them as complementary (i.e. their combinations are useful and insightful)? As indicated above, the way you view reality will influence other aspects of your research design.

Following Rossman and Wilson (1985), researchers can be grouped into three categories, depending on their views on reality. One category of researchers holds the view that the objective and subjective perspectives on reality are mutually exclusive. In other words, the root assumptions on which each of the two perspectives is grounded are very different and must not be blended. Researchers adopting this viewpoint are labelled *purists* in the literature.

Another category of researchers adopt a flexible attitude to the two perspectives. These researchers argue that all social phenomena have many sides and interpretations. It is therefore useful to combine both objective and subjective perspectives on reality in a single research thesis in order to gain a deeper insight into the research issue that is being studied. In other words, the objective and subjective perspectives are best seen as points on a continuum rather than as alternative perspectives. But the degree of emphasis on objective or subjective dimensions of reality must depend on every research situation – i.e. how much insight can be gained in using the various approaches or their combinations. In some research situations, researchers may lean more on objectivist views of reality, in some situations, the subjective views may be preferable and in other situations, some combination of objectivist and subjectivist views of reality may be appropriate. Researchers endorsing this view are labelled *situationalists*.

The third category of researchers holds the view that the nature of research issues and the objectives of an investigation should determine the view of reality that a researcher adopts. These researchers are labelled *pragmatists*. They neither accept nor reject the notion that the two perspectives on reality can be combined or used separately. They simply say that the nature of the research task rather than the research situation must decide what view of reality you may consider to be appropriate. Some tasks are suitable for objectivist view of reality while other tasks may be suitable for subjectivist view of reality. If your problem formulation shows that an objective or a subjective perspective of reality is appropriate you must choose what is appropriate. Furthermore, you may also combine two or more views of reality in a single thesis if the problem formulation shows that a combination of views of reality is appropriate in providing the best insight.

One can say that there are sheer nuances of difference between the situationalist and the pragmatist researcher. But while the situationalist may be more inclined to using a combination of perspectives in a given thesis, the pragmatist is more likely to select one perspective, but does not reject a combination. The purist, on the other hand, rejects combinations under all circumstances.

The discussions above suggest that you need to state explicitly if your understanding of reality (and how to apply it in research) follows the views of a purist, a situationalist or a pragmatist. You must also provide arguments that justify your views and discuss them with your supervisor.

CHAPTER SIX

THE ROLE OF THEORIES IN YOUR THESIS

Introduction

An important requirement in your thesis work is that you must demonstrate a good understanding of what other people have written about the issues central to your own investigation. That is, your thesis must be grounded in existing knowledge about your subject. To do so, you need to review the existing literature that is relevant to your problem statements. The literature review forms the theoretical foundation of your study.

Some students find this literature review to be unnecessary, particularly if they are much more interested in solving what they consider to be mainly a "practical" business, social or institutional problem.

The aim of this chapter is therefore to emphasise the importance of theories and to draw your attention to the various types of theories that exist in the social science literature. It also discusses the roles that theories play in solving social science problems and provides you with some guidelines on how to conduct your literature review.

What is a Theory?

The meaning of the word "theory" varies with the context in which it is used. In social science theories may be defined as series of systematic inter-related statements or generalisations that explain and/or anticipate developments in a specific context or phenomenon. Strauss and Corbin (1998:15) define theory as "a set of well-developed concepts related through statements of relationship which together constitute an integrated framework that can be used to explain or predict phenomena". In other words, theories form the basis of frameworks for conceptualizing and organizing your thoughts about the issues that you are concerned with in your research. That is, they provide the language; the concepts and assumptions that will help you make sense of the phenomenon that you seek to investigate.

I find Deutsch and Krauss' (1965: vii) viewpoint on theories to be a useful way of understanding their role in project work. They describe theories as a net that researchers weave "to catch the world of observation – to explain, predict, and influence it. The theorists . . . have woven nets of different sorts. Some are all-purpose nets meant to catch many species of behavior, others are clearly limited to a few species; some have been constructed for use near the surface, others work best in the depths".

Thus, theories enable you to connect the issues you are investigating to the existing body of knowledge in the area.

Classification of Theories

There are different classifications of theories in the social science literature. One of the well-known classifications identifies the following four levels of theory:

1. Metatheories
2. Grand theories
3. Mid-range theories
4. Micro theories

Metatheories describe the broad philosophical assumptions concerning reality that are accepted in social science as clearly demarcated boundaries of thoughts in a particular field of study. An adoption of a metatheoretical position in a particular research therefore implies commitment of the researcher to conceptual assumptions underlying that metatheory. For example, cultural studies are usually based on different metatheoretical assumptions. Some scholars conceive culture as a component of a social system aimed at fulfilling specific social functions (i.e. a functionalist perspective of culture). Others see it as a product of historical evolution (i.e. social diffusionist perspective). Yet others see it as outcomes of human cognition (i.e. ideational perspective). See Allaire and Firsirotu (1984) for an elaboration.

As indicated in chapter 5, these assumptions are embedded in the ontological, epistemological and methodological views of the researcher. In other words, meta-theories serve the following purposes:

1. Clarify the general assumptions underlying a subject matter
2. Specify the important problems faced in undertaking investigations, and
3. Specify what are acceptable methods

The concept of metatheories is frequently used interchangeably with the concept of paradigm.

A *grand theory* is defined as an all-inclusive unified theory that seeks to explain social behaviour, social organization and social change in human experience. It normally provides the key concepts and principles of the social science discipline and is therefore consistent with the dominant meta-theories or paradigms of the discipline. For example, grand theory in economics defines the laws of scarcity and needs as well as the relationship between demand, supply and prices. Feminism, Marxism, and Democracy are also grand theories in political science with cross-disciplinary implications for other social science disciplines.

Grand theories may also be found in specific fields of study. For example one grand theory in cultural studies is that there are universally identifiable dimensions of culture that can be found in all societies and groups. Cultures can therefore be compared on the basis of these universal dimensions and the manner in which they evolve over time in the history of a society (see Hofstede, 2001). An alternative grand theory in culture is that culture is socially constructed and continuously negotiated among groups of people. As such it cannot be studied by using universal dimensions (see Gullestrup, 2006)

Mid-range theory is a term that emerged in sociological studies in the 1940s (frequently associated with the works of Robert K. Merton). It represents theories that connect grand theories with empirical evidence. It consists of limited sets of assumptions from which specific hypotheses are logically derived and confirmed by empirical investigation (see Merton, 1968). Thus, when you engage in literature review, you are mostly discussing mid-range theories. For example, if you adopt a "universalist" approach to cultural studies, you will review the works of scholars that have adopted this approach in specific empirical investigations and use their findings to justify your hypothesis formulation. Similarly, if you adopt a social constructivist approach to your cultural research, you will justify your arguments by using studies that have adopted this approach.

50

Micro theories constitute the lowest level of theories. They focus on individuals or small groups located in specific contexts. As such, explanations found in micro theories are of limited generalization on their own. They can, however, constitute essential inputs in the generation of new perspectives and theory development. Some scholars use case studies to generate micro theories that are then further developed through multiple case studies to become important inputs in mid-range theory formulation.

Again, continuing with the example with cultural studies, you may undertake an investigation of the culture of a sports organization (e.g. a football club) in a particular community. You may gain a good understanding of how the culture of that club has developed over time and how it has influenced the accepted rules of behaviour in the club. Although this investigation cannot be generalized across all football clubs in a particular country, it can produce useful inputs into how football clubs in small communities may be investigated.

Your awareness of the differences in the levels of theory informs your readers that you understand the nuances in the research you are doing. This also helps in positioning your own work within the existing body of work in the field.

As students you can aspire to develop mid-range theories – e.g. as a way of providing general statements from your empirical investigations, if research designs enable you to do so. In doing so you will be contributing to the extension of the existing boundaries of your specific fields of research and at the same time, deepen your understandings of the core content of the subjects that you are studying.

From Micro to Mid-Range Theories

Theories that are originally formulated at micro levels of analysis may be adopted by other scholars in their own studies. As an increasing number of researchers find the theory useful, it becomes gradually "upgraded" from its micro levels to mid-range theory level. For example, a student may find it useful to distinguish between "official" and "unofficial" costs in an investment analysis in Africa, although these concepts are not commonly used in the literature. He may use the term official costs to describe what is legally defined to be recorded as costs. But unofficial costs may be what some may popularly describe as bribes, which are nonetheless essential for any effective running of a business in Africa. He may also distinguish between primary level unofficial costs and secondary level unofficial costs. Primary level unofficial costs may be those found at top management/administrative and official circles while the secondary level unofficial cost may represent lower level briberies that are given to junior officers to speed up work processes within bureaucratic circles. While secondary level unofficial costs may be effectively absorbed under impress, the primary level unofficial costs may be too big to be covered by impress.

Other scholars may find this distinction as insightful and begin to adopt it. Over time, this mode of conceptualization will become an accepted construct in the business economics literature.

Use of Theories

Theories are found in the literature that you read – i.e. in the books and articles written about the subject that you choose to write about. Each of the levels of theory listed above will play a different role in your thesis. The meta-theories define the philosophical foundations of your thesis. The grand theories define the boundaries of your subject of investigation. They combine with the meta-theories to establish the platform on which you will base your research. But much of the discussions in your theoretical chapters will draw on mid-range theories. Through studying previous work done on your topic, you will become acquainted with the mainstream concepts used in the literature to describe your problem and its constituent elements. Thus, when you do your literature review you are mostly engaged in a discussion of mid-range theories.

Mid-range theories can also be used to develop an *analytical framework* – i.e. variables to analyse and the connections between them. They therefore help you identify what kind of information you require for the analysis and what is the most appropriate means of acquiring this information. Without such an analytical framework you will risk drowning in a sea of information, since you will be unable to sort out the relevant from the irrelevant information.

There are, however, many competing theories within each subject area. An essential task in the thesis work process is for you to discuss most of the leading theories that attempt to explain the problem of interest, comparing the strengths of the arguments underlying them and their empirical foundation. In this process of discussion, it is important to draw on the views of other writers in the field who have undertaken similar investigations. Such discussions will sharpen your understanding of the theories as well as their limitations in explaining the problems you are investigating.

All the theories that you use in your research need not be discussed in a single chapter. If you decide to use multiple sets of theories, each set can be discussed in a separate chapter. You can then write another separate chapter to synthesise the discussions in the preceding theoretical chapters and use this chapter to present your overall analytical framework for the thesis.

Your theoretical chapters will become richer when you discuss the theoretical tensions or oppositions in the existing literature. This will help you to identify theoretical gaps and justify new perspectives.

Guidelines for Literature Review

It is important for you to read the most recent literature that reflects the most recent knowledge on the topic you choose to write on. It is also important to read published studies in journals in addition to textbooks. Box 6.1 provides a list of questions that can guide you in your review of the existing literature.

Box 6.1
Guidelines for Literature Review

- What are the main concepts and ideas in the book or article that I have read?

- Do the author's ideas agree with or corroborate ideas I have come across in other articles or books? Or do the ideas differ from them? What are the points of agreement and/or differences?

- How do I explain the differences in the viewpoints?

- Is there any agreement among the authors on the definition of concepts and angles of perception?

- Which of these concepts, ideas and conclusions do I find relevant to my own work and why?

In reviewing the literature three main factors need to be carefully examined:

- The date of publication and the years in which the researcher has collected the data used.
- The country in which the study has been conducted (if it is an empirical piece of work).
- The meta-theoretical approach adopted by the researcher and the data collection methods used.

Remember that what you read from previous studies is usually a snapshot of the phenomenon that other researchers have investigated – i.e. what they have found in a given context and at a given point in time. Since you are likely to use the knowledge in that study within a different time frame, it is important for you to examine the validity and relevance of the knowledge at the time of the publication of the article or book that you have read.

The research setting (i.e. the country within which the writer has collected his/her data) is also important to note since it provides an idea about the context of the knowledge presented. For example, a study done about export promotion in Ghana may not necessarily reflect requirements for export sector development in another African country.

CHAPTER SEVEN

QUALITATIVE DATA COLLECTION METHODS AND TECHNIQUES

Introduction

Qualitative research methods have increased in popularity among social science researchers during the last three decades. Protagonists of qualitative research methods emphasize their ability to provide complex textual descriptions of how people experience a given research issue. In other words, they are highly useful in providing information about the "human" side of an issue – i.e. producing rich insights into the often contradictory behaviours, beliefs, opinions, emotions, and relationships of individuals in specific situations.

This chapter introduces you to the general characteristics of these methods and techniques. It also draws your attention to some of the challenges you should be aware of when you decide to use them, and what you should do to strengthen the quality of your work.

General Characteristics of Qualitative Methods

The term *qualitative method* is generally used to represent a wide variety of data collection methods. These include *ethnography, participant observation, in-depth interviewing,* and *conversational interviewing* (Bryman and Bell, 2011). Strauss and Corbin (1998: 10-11) define qualitative research as "any type of research that produces findings not arrived at by statistical procedures or other means of quantification". Such types of research are described as emphasizing "cases and contexts". That is, they engage in detailed examination of cases that are related to their chosen topics and present "authentic interpretations that are sensitive to specific social-historical contexts" (Neuman 2006:151).

If you choose to use qualitative data collection methods you must design your research in a way that enables you to get a first-hand look at the settings in which those you study operate and see what the participants describe in their answers. Qualitative methods also allow the participants to raise topics and issues which you may not anticipate and which might be critical to the investigation. Furthermore, they allow participants to express their feelings and offer their perspectives in their own words. In other words, if your study is concerned with gaining newer insights into the phenomenon that you are investigating rather than finding confirmation for existing theories, you must strongly consider using qualitative methods to collect your data. Qualitative studies may therefore be useful in developing micro theories.

Qualitative Data Collection Techniques

Focus Group

The focus group technique allows you to bring a selected number of people together to discuss the issues that your investigation centres on. For example, if a company is about to introduce a new product into a particular market and is not sure about how consumers will respond to the product and marketing strategies, the company may ask you to conduct a focus group study to gain insights into potential consumer perception. To do this you will select about 12 individuals who do not know each other in advance. Your selection should be guided by some specific criteria (e.g. age, income, and levels of education) that are of particular interest to the company. You then arrange with them to meet and discuss various aspects of the product – quality, colour, size, price range, and typical shops in which such a product will be expected to be sold. The session typically lasts for one to two hours. You will act as a facilitator of the discussion. But the participants will be required to engage in a free discussion of the issues that you consider important for your investigation. The free discussions allow them to explore and clarify their views in ways that would be less easily accessible in one-to-one interviews. As a facilitator, you are expected to ask probing questions, when necessary in order to stimulate the discussion. You will usually bring someone along to take notes while the discussion goes on. The sessions may also be video-taped if the participants agree to such an arrangement. When group dynamics work well focus group session will provide new and unexpected perspective on the marketing strategies that the company might have been contemplating on adopting.

The approach has been found to be highly useful in non-business related studies where the research seeks to to explore themes that are not well-known to him or her. For example community health studies have increasingly adopted focus group studies in examining issues such as locally held beliefs on the value of immunization or traditional practices in rural communities.

Focus group studies have also been found to be a useful supplement to quantitative methods. It enables researchers to generate new hypotheses; to explore intermediate variables as a means of explaining certain relationships found in survey data; or to validate findings gathered through other methods using triangulation for comparison of different perspectives.

Observations

When you are dealing with a subject that people are likely to feel uncomfortable or unwilling to discuss you may want to use observations as part of your data collection technique. You may use your observations simply to describe the phenomenon – i.e. write down what you observe – or to make inferences about what you observe, or even make personal assessments. For example, you may be able to observe what people are doing and through these observations make inferences about their attitude, based on certain representations of what you observe is a typical characteristic. But since you may be wrong in your inferences and evaluations, it is a common practice to engage in dialogue with those you observe (i.e. qualitative interviews with those under observation) in order to check the accuracy of your inferences. Observation studies are of two basic types – **non-participant** and **participant** observations. If you are a non-participant observer, it means that you visit the people you observe but do not take part in their daily activities. This makes you a "stranger" to the environment and therefore you "intrude" on the situation that you observe. Your presence may therefore make the observed situation less normal, since the participants will be aware that they are being observed and alter their behaviour accordingly. Such a technique may be necessary, if you (as an observer) do not have competencies required to participate in the daily work process of people you observe. For example, if you are observing nurses at work, you may not be able to participate directly in their actual nursing process unless you have the training and competencies to do so.

Participant observation studies, on the other hand, require you to become an active participant in the environment in which the study is conducted. You will do exactly what those you observe do and record your observations in the process. You will usually engage in informal conversations with other participants as part of the data collection process.

When you use a participant observation method, you will need to spend several days (and maybe months) with the participants in the research thesis. For example, if you are doing an investigation in a company, you will need to observe the employees of the company while they are working, engage in small talks with them, take notes of what you hear in the corridors and canteens, etc. If you are invited to sit in at some of their meetings, this will provide you with a good opportunity to gain first-hand impression of how decisions are taken. You will also be allowed to supplement the information you get from these observations and conversations with written documents from the company and formal interviews with some key employees.

The advantage of using the participant observation method is that it gives you a profound understanding of the setting within which the research is done. This will enable you to provide a description of the context within which the participants operate.

The most obvious difficulty with participant observation is that you cannot avoid becoming involved with the group at a social and emotional level. This may well influence your own behaviour and change the very phenomena under study. That is, participating in the group activities may make an unbiased collection of data difficult. You may be able to reduce the bias when you deliberately record your observations frequently and reflect on what you observe. You must not rely too much on your memory – i.e. record your observations only at the end of the day or week.

If you serve an internship with an organization, participant observation becomes one of the useful data collection techniques you must consider. The same holds true for those of you doing your studies on a part-time basis while working. That is, if your research is done in a company where you work, the participant observation method will definitely be a useful data collection method you must consider.

Remember that it takes a long time and a lot of effort to collect data using observational methods. Participants' behaviours must be observed over a long period of time for you to be able to make reliable inferences. Furthermore, as a researcher, you cannot be everywhere at all times. What you observe is therefore a snapshot and partial evidence of what actually happens. You must therefore be careful in making generalizations on the basis of your observations.

Qualitative Interviews

Interviewing is another technique you can use to collect qualitative data. As we can see in chapter 9, interviewing is used in collecting quantitative data as well. But there are differences. Qualitative interviewing seeks to gain an insight into the "lived-experiences" of the person you are interviewing. It provides you with the opportunity to listen to what respondents themselves say about issues that you investigate in their own words. It can be used alone in collecting the required data, or in conjunction with other qualitative data collection techniques.

One of the most popular techniques in qualitative interviewing is *the critical incident technique* (CIT). This is a useful technique in gaining insights into people's experiences and the impact of such experiences on their perceptions and behaviours. Flanagan (1954:333) described the critical incident technique as "an observable human activity that is sufficiently complete in itself to permit inferences and predictions to be made about the person performing the act". The technique is composed of two facets: (1) the critical incident itself, and (2) reflections. The critical incident may be a snapshot, a situation or an encounter which the person has been engaged in. The reflective component involves engaging with and exploring the incident with the person that you are interviewing on both intellectual and emotional levels. The aim of the reflection is to reach a new understanding of the experience with the person.

Thus CIT allows the people you interview to describe freely their experiences and unreservedly express their feelings and to reflect on their experiences while they are talking to you. In this way you and the respondent will be able to explore new dimensions in your investigation.

You will usually start the process by asking the respondent to describe some critical incidents which he experienced personally in the field of activity being analysed. He then narrates the event naturally just as he would do in a conversation with friends or acquaintances. He may not remember the events in chronological order and may go back and forth in narrating his experiences. Allow him to do so, and take notes as he speaks. You may tape the interview if the respondent allows you to do so.

Such an interview may take several hours at one go or several days with short interviews per day. But in the end, you will come out with a richer understanding of the person's lived experience and gain more insight into the subject of your investigation.

Researchers are advised to organize the data they collect into files or notebooks with each critical incident being coded with a unique number. Incidents are then carefully read and sorted into content themes in an iterative process. This consists of repeated careful readings of events in order to categorize reported experiences into similar categories.

Challenges in Collecting Qualitative Data

Those of you who consider statistical methods to be difficult may want to choose qualitative research method in order to avoid the statistics. Others may consider qualitative methods to be a lot easier and less time-consuming. You will be making mistakes in basing your choice on these motives. In fact, qualitative methods are difficult to use. As noted above, the methods require you to seek an understanding of the social phenomena you investigate through the voices of those you investigate. In its pure form, it means that you should avoid imposing preconceived ideas on your data – i.e. you would not usually start your research with a theoretical understanding of the phenomenon you seek to analyse. Instead you will spend much time on your data analysis and let your understanding emerge from the data. This approach is referred to as *inductive research* in the literature (Denzin and Lincoln, 2003). Students are, however, not required to engage in "pure" inductive research when writing their theses. They are expected to do a literature review in connection with their thesis writing. But the theories they discuss under their literature review will basically provide them with a *pre-understanding* of the issues they seek to investigate (see Bryman and Bell, 2011). They will then use the qualitative data they collect to analyse the extent to which the existing theories explain the issues they investigate and at the same time produce new insights that emerge from the data. Thus, getting good results from qualitative data requires substantial skill and experience.

Your personality and the personality of those you study also have some influence on the data you are able to collect. The personalities may influence the extent to which deeper meanings and feelings are explored and revealed through interviews. Some of you may be hesitant to press further for richer insights. In some situations people you interview may be guarded in their responses.

There is also the question of the validity of your investigations. How can you be sure that you have not misunderstood what you have observed or have provided a misinterpretation of what you have been told by your respondents? Or can you be absolutely sure that those you interview are honest in their narrations?

In many qualitative research situations, interview responses may represent "edited" versions of realities that are purposefully presented by participants. They therefore reflect individual sense making of their roles as part of their overall image construction. The challenge is how researchers can make sense of the respondents' intentions with the responses that are given.

It is also important to bear in mind that qualitative data collection processes can lead the researchers on to circumstances that may seriously test his ethical standards. For example, he may stumble on evidence bordering on the verge of illegality or moral abhorrence. Or respondents who repose trust in the researcher after months of close interaction may make statements "off-the-record", but which on closer reflection may be found to be an important piece in the puzzle that the researcher has been grappling with. The ethical challenge lies in what the research decides to do with such data.

Issues of Data Size

Immediately you start thinking of how to collect your data, you must begin to think about how the data is going to be analysed. You must start organizing your data from an early stage in your research process. If you collect more data than you have the necessary time and other resources for analysing, you are wasting time and effort. On the other hand, if you do not collect enough good data to address your research questions adequately you may realise that it may be too late to go back and supplement the data that you have collected. The quality of your research will then be compromised. Box 7.1 provides you with some guidelines on how to address this challenge.

Box 7.1
Guidelines for Deciding on Data Size

- What kinds of research data will I (need to) collect?
- How will this data help/allow me to address my research purposes?
- How will I collect this data?
- How much data will I collect?
- How will I validate this data or establish that it is good quality data?
- How will I organize this data so it will be in good shape for being analysed?
- What forms of data analysis will I use?
- What justifies these forms of analysis? How will they help me achieve my research purposes?
- How are these forms of analysis conducted?
- What do I need to know in order to use these forms of analysis in an expert way?

The question is always asked as to whether it is scientifically legitimate to alter and even add collection methods during a study. Opinions differ widely on the issue. Some scholars argue that inconsistencies in data collection method during a piece of research reduce the comparability of the data across settings. Others encourage researchers not to shy away from using new data collection opportunities as and when they arise.

My experience with this type of research tells me that it is advisable to collect the data in batches. I will usually analyze a small data batch carefully see how far it takes me in my understanding of the phenomenon and only then determine what additional data I will need.

Evaluation of Qualitative Studies

The discussions above suggest that you must give your readers a complete picture of the data collection and analysis process. You must not let your readers guess how you have done what you have done – i.e. they must not be left at the mercy of your intuitions. This raises the issue of evaluation. Qualitative studies are usually evaluated on the basis of (1) *trustworthiness* and *authenticity* (Silverman, 2010; Bryman and Bell, 2011).

Trustworthiness is assessed in terms of the following dimensions:

Credibility: This examines the extent to which you have followed the accepted procedures in conducting qualitative investigations. Usually, you must send your interview transcripts to your respondents for them to confirm that you have correctly understood what they have told you. In other words, your research will achieve greater credibility when your respondents have validated the data you have collected. This is referred to as *respondent validation*.

Transferability: This requires you to provide a detailed account of the context within which your study has been conducted. This will enable future researchers to compare your study with theirs in order to determine whether your findings hold true in other contexts.

Dependability: This requires that you keep detailed records of all phases of the research process – problem formulation, selection of research participants, field work notes, interview transcripts, etc. These materials will provide evidence that you have done the study in the prescribed manner. In other words, the dependability criterion reinforces credibility and transferability criteria.

Confirmability: This requirement adds further weight to the three previous criteria of trustworthiness. It requires you to demonstrate that you have acted in good faith all along in the research process. In other words, you do not have any other interest in the research than to understand the reality that you set out to investigate.

The second evaluation criterion, *authenticity*, relates to the extent to which your investigations are fair (i.e. include all relevant people and their viewpoints), improve understanding of the social phenomenon that you seek to investigate, and provide opportunities for those involved to improve their insight into their own situations and act to change them, if they deem it necessary to do so.

Approaches to Qualitative Data Analysis

There are a number of methods that you can use in analysing qualitative data. But you need to read a handful of reports that have adopted these methods to be able to gain some insight into how to use them and to understand theories and assumptions underlying them. I provide an overview of some of them in this section. The aim is to stimulate your interest in them and to encourage you to read further about them. The methods described here can be used individually or in combinations in order to produce richer insight from the dataset that you have at your disposal.

Grounded Theory

This is an approach to building theory based on available data collected in a variety of ways. The research principle behind grounded theory method may combine both inductive and deductive approaches. It sees data collection, data analysis and theory development as an iterative process that must be repeated until the researcher is satisfied with the description and explanation of the phenomenon being investigated. Thus, if done well, the resulting theory will fit the dataset quite well and provide convincing insights into the phenomenon.

The most common approach to the data analysis is to read (and re-read) a textual database and "discover" or label variables (called categories, concepts and properties) and their interrelationships.

Phenomenology

In simple terms phenomenology is the study of phenomena: their nature and meanings. A researcher that uses this approach focuses on the way things appear to people through experience or in their consciousness. The phenomenological researcher therefore seeks to provide rich descriptions of "lived experiences". The underlying reasoning is that human beings live their lives in a world that is filled with complex meanings which form the backdrop of people's daily actions and interactions. The task of the researcher is to understand what guides human interactions and to use these dimensions to describe and explain the way individuals live and make sense of their world in their particular ways.

The challenge that you are likely to face as a phenomenological researcher is how to help those you investigate (the participants in your research) express their world as directly as possible. You will also face the challenge of using the appropriate words to reveal the experiences of the participants to your readers.

The most common methods used include interviews, conversations, diaries, participant observation, action research, focus meetings and analysis of personal texts.

Discourse analysis

Discourse analysis relates largely to the interpretation of text – previous writings or transcripts written by the researcher or his/her colleagues. It involves being critically attentive to patterns of language in use in describing an event or experience and the circumstances (participants, situations, purposes, outcomes) with which they are associated. This is done deliberately, systematically, and, as far as possible, objectively, and to produce accounts (descriptions, interpretations, explanations) of investigations. As a method of data analysis, discourse analysis is part of applied linguistics.

CHAPTER EIGHT

QUANTITATIVE DATA COLLECTION METHODS

Introduction

If you seek to test specific hypotheses or find numerical answers to specific elements in your research questions, you may strongly consider using quantitative data collection methods. There is a variety of quantitative data collection techniques that you may consider. But this chapter introduces you to the two most popular ones – questionnaire-based surveys and interviews. The chapter also provides you with an overview of the general characteristics of quantitative data collection methods and guides you in writing good questionnaires and conducting interviews to gather quantitative data.

General Characteristics of Quantitative Methods

Quantitative data collection methods allow you to test hypotheses derived from theories you have read about the issues you are investigating in your thesis. Such studies will usually encourage you to investigate causal relationships between specified variables. In other words, your theories will indicate that some specific variables influence other variables to produce an effect.

Let us take an example of a study of youth performance in schools in a given region in Ghana. Let us assume that you are interested in understanding why some young people within the age groups of 15 and 20 in district "A" do less well in school than young people similar age groups in district "B" in Ghana.

You will need to identify which variables have been found in previous studies (e.g. in other regions) to influence young people's performance in classes. You will then want to test whether the previously identified variables impact performance of the young people in the district in which your study is conducted.

Let us take another example. If you are interested in differences in the performance of companies in Ghana, you will have to identify measurable indicators of company performance (e.g. profitability or market growth rate) and identify factors that impact company performance. You will then collect quantitative data from a number of companies to determine whether the relationships identified in previous studies hold true in your study as well.

Quantitative methods are generally less flexible than qualitative methods. There are standardized procedures and techniques for collecting, organizing and analysing the data. These standardized and well accepted procedures tend to accord quantitative methods with "a scientific" image and to make them very popular in social science research. The systematic and standardized data collection procedures also allow you to collect data that are sufficiently general and make the results of your investigation generalizable.

Steps in Quantitative Data Collection

Steps in the survey research process are outlined in Figure 8. 1. They start with defining the survey objectives, and continue with developing a sample frame, specifying the strategy for data collection, and conducting the appropriate analyses as well as evaluation. Each of these steps is critical to the success of the survey. You must therefore take a holistic approach to the survey design by consciously considering all aspects of the survey process.

The steps outlined here build on the steps in the overall thesis work process presented in chapter 3 (see Figure 3.1). Your definition of the survey objectives in the data collection phase must be consistent with your problem formulation as well as the theories, concepts and models that you have discussed earlier in the thesis. You will also derive your hypotheses from your literature review and theoretical discussions.

Figure 8.1: **Steps in the Process of Survey Research**

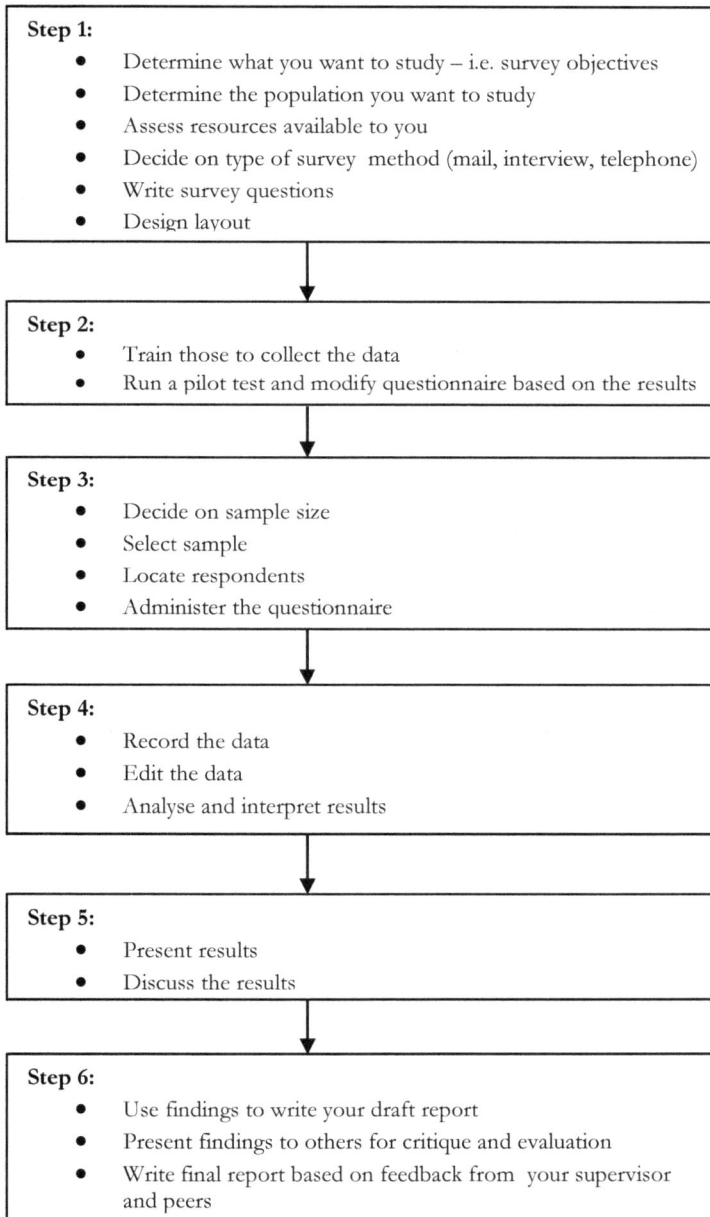

Step 1:
- Determine what you want to study – i.e. survey objectives
- Determine the population you want to study
- Assess resources available to you
- Decide on type of survey method (mail, interview, telephone)
- Write survey questions
- Design layout

Step 2:
- Train those to collect the data
- Run a pilot test and modify questionnaire based on the results

Step 3:
- Decide on sample size
- Select sample
- Locate respondents
- Administer the questionnaire

Step 4:
- Record the data
- Edit the data
- Analyse and interpret results

Step 5:
- Present results
- Discuss the results

Step 6:
- Use findings to write your draft report
- Present findings to others for critique and evaluation
- Write final report based on feedback from your supervisor and peers

Source: Based on multiple sources Neuman (2006), Bryman and Bell (2011)

Quantitative Data Collection Techniques

You can use different techniques in the data collection process. The main ones include surveys and interviews.

Questionnaire is the most conventional data collection instrument used in surveys. Writing questionnaires is, however, not as easy as you may think. It requires practice, patience and creativity. You must make sure that respondents perceive the questions to be clear, relevant and meaningful. The same words may convey different meanings to different respondents. This is particularly a problem in cross-national research situations where respondents are likely to have different frames of reference. Furthermore, the fact that respondents participate in surveys out of kindness rather than economic rewards means that questionnaires must be attractive for them to bother to fill and return them. Thus, without treating questionnaire design in a professional manner you will risk a low response rate. Thus, the overall requirement in questionnaire design is that it must be respondent-friendly. Box 8.1 provides you with some guidelines on writing good questionnaires.

Box 8. 1
Guidelines for Writing Good Questionnaires

1. You may structure your questionnaire as follows:
    Section 1: Information to fulfil basic objectives of the study
    Section 2: Information to fulfil other research objectives
    Section 3: Identification (i.e. demographic data of the respondent)
    Section 4: Thank you statement

2. Your initial questions should be captivating, general, clearly stated and easy to answer. Place difficult questions towards the end of the questionnaire

3. All questions on a particular topic should be grouped together and in a logical fashion before the questionnaire moves on to another topic.

4. Keep your questions brief and the questionnaire itself must be short.

5. Make sure that there are no misspellings or grammatical errors in the questionnaire

6. Avoid controversial, embarrassing, or emotionally charged questions as much as possible. If these are absolutely necessary, place them towards the end of the questionnaire.

7. Consider placing lifestyle questions toward the end of the questionnaire. Some respondents are uncomfortable with such questions, but are willing to provide the information after going through the other questions.

8. Avoid double-barrelled questions.
    A double-barrelled question is the type that requires the respondent to address more than one issue at a time.

9. Avoid leading or loaded questions.
    A leading or loaded question is the type that directs the respondent to give a specific answer.

Sources: Based on Shao (1999); Neuman 2006

There are two types of questions that you may ask respondents to answer – closed and open-ended questions. A closed question is that which can be answered with either a single word or a short phrase. Examples include questions seeking demographic information such as age, sex, levels of education and marital status of respondents. Closed questions are easy to answer and can be answered quickly.

If your respondents are not well educated, it may be a good idea to make most of the questions closed since this will make it easier for them to complete. It will also be easier for you to code and analyse closed questions than open-ended ones.

Open-ended questions normally demand long answers. Some of them require the respondent to think and reflect on specific situations or events before answering them. In this way, respondents may provide more nuanced insights into the phenomenon you seek to investigate. On the other hand, open-ended questions are more difficult to code and analyse.

Your questionnaire must be accompanied by a covering letter that introduces the potential respondent to the research thesis, stresses its legitimacy, and encourages participation. When you plan to administer the questionnaire to the respondent yourself, mail a covering letter to the respondents prior to the face-to-face contact. This will help you break the ice and create a good atmosphere for the interaction during your meeting with the respondent. Remember the following points when writing your covering letter:

1. Address letter to the specific prospective respondent.
2. Use your university's professional letter-head stationery.
3. Specify the general topic on which you are conducting your investigation and stress its importance to the prospective respondent.
4. Give assurance that the prospective respondent's name will not be revealed.

5. Communicate the overall time frame of the study to the respondent to solicit his/her involvement. This should also include the completion date for the questionnaire.
6. Communicate where and how to return the questionnaire.
7. Provide advance thank-you statement for willingness to participate.

Administering Questionnaires

There are various methods of administering questionnaires. These include mail, online or face-to-face presentation of the questionnaire to the respondent by the researcher himself. The factors you need to take into account in deciding on the appropriate method include the objectives of the study, the sample and its geographical distribution, the types of questions, the resources at your disposal to do the research as well as cultural and behavioural tendencies among the respondents.

Administering through Mails

Mail surveys are self-administered questionnaires, i.e. the respondent fills in the questionnaire himself. By using mail survey technique, your respondents (i.e. those who provide you with the data) can do so without revealing their identity – i.e. by answering the questions anonymously. This will allow them to respond with openness and to sensitive questions. Mail surveys also avoid interviewer-bias and are relatively cheap compared to interviews. You do not have to arrange meetings with each individual respondent. Thus, if you have a limited research budget and want to collect data from many respondents, you must seriously consider mail survey. But in countries where mail systems are not very reliable, this technique may be less appropriate.

Furthermore, if respondents are not particularly interested in the subject matter, most of them may fail to respond and the survey may suffer from low response rate. Further, mail survey approach denies the analyst the rich information that may be gathered through interviews (allowing respondents to clarify their answers) and observations.

Online Surveys

The development and pervasiveness of web technology in the developed countries now provides a cheaper and speedy technique of conducting surveys. Developing countries are also catching up very fast in the use of online surveys. Estimates show that the use of online survey technique can reduce costs by 20 to 40 per cent and provide the results in half the time it takes to do traditional mail surveys. Furthermore, if you use online survey technique, you can increase your sample sizes without corresponding cost increases. For example, those groups of professionals (e.g. doctors and top level executives) who find it difficult to allocate time to interviews may be willing to respond to online surveys that they consider relevant but at their own convenience. In addition to this, it will be easier for you to send reminders to respondents through electronic mails and time them to arrive during the less busy time of the day. You can also more readily analyse the results of online surveys.

Quantitative Interviews

We noted in chapter seven that you can collect qualitative data through interviews. You can also collect quantitative data using interviews. Interviews can be conducted in two main ways; either through face-to-face interactions or through telephone or video-based interactions. The main advantage of telephone interview is that the respondent can give his/her answers anonymously.

The first step in an interviewing process is respondent selection. That is, whom should you talk to? If you are collecting data from companies and institutions, you can select your respondents in a combination of two ways: (1) as representative of an organisation, and/or (2) as private individuals, i.e. in their personal capacity. The mode in which respondents are selected influences the content of their response. Respondents who are selected as representatives of their respective organisations tend to speak *for* their organisations, i.e. represent an official or "public" face. Respondents who are selected in their individual capacities speak *of* the organisations and on other management issues in general, and may tend to be more critical and open in their views.

The manner in which questions are asked during an interview process will determine whether the respondents reply either *for* their organisations or talk *of* the organisations. The interviewer must therefore be mindful of the changes in orientation made by the respondent during the interviewing process.

Another important problem faced in the respondent selection process is the potential censorship built into the process. This relates to the requirement in many organisations that top managers must approve of the respondents selected. There is, therefore, the danger of indirect data censorship in the sense that management may restrict the coverage in order not to expose what they consider to be bad sides of the organisation.

Selection bias may, however, be an outcome of necessity. It is natural to expect that the most detailed information is obtained from those respondents who are willing to share their thoughts and feelings with the interviewer about different situations within the organisation and have the time to do so in details. It is always difficult to find many of those people within an organisation. It is therefore not unusual for interviewers to select their respondents on the recommendations of those they have earlier interviewed. In situations where respondents are selected through recommendations of previously interviewed respondents there is a danger that the chain of selection becomes friends of the initial respondents or friends of their friends. In this way, the sample becomes purposefully selected rather than reflective of the entire spectrum of the organisation. This can weaken the validity of the accounts they give as representing an overall perspective of the organisation.

Types of Interviews

Interviews are classified into two groups: (1) standardised interviews, and (2) non-standardised interviews. Non-standardised interviews may be either semi-structured or in-depth interviews. *Standardised interviews* are normally used to gather data which will then be the subject of quantitative analysis. Non-standardised interviews serve the purpose of gathering data for qualitative analyses. As Saunders *et al*

(2007:313) explain it, "these data are likely to be used not only to reveal and understand the 'what' and the 'how' but also to place more emphasis on exploring the 'why'.

Standardised interviews use a questionnaire-type format. Some scholars refer to them as *interview-administered questionnaires* or *quantitative research interviews* to indicate that they are largely used in connection with surveys. The interviewer reads each question to the respondent and then records his responses on a standardised schedule, usually with pre-coded answers. Structured interviews have all the advantages of questionnaires. They also carry the benefit of the interviewer clarifying any question on the list that the respondent may have difficulties in understanding. Interviews also provide respondents with the opportunity to reflect on events without needing to write anything down. The face-to-face interaction with an interviewer also gives the respondent some degree of confidence to divulge sensitive information. Not many people are willing to fill questionnaires and provide sensitive information to people that they have never met, even if assurances have been given by the researcher that the responses will be treated with anonymity.

In *semi-structured* interviews the researcher will have a list of themes and questions to be covered. The questions may, however, vary from one interview situation to another. The order of questions may also be varied depending on the flow of the conversation. The variations allow the researcher to introduce additional questions where relevant in order to explore specific dimensions of the research question in specific interview situations – e.g. where the respondent has special or expert knowledge to share on specific issues.

Unstructured interviews are a lot more informal than the semi-structured interviews. Researchers use them to gain deeper insights into general areas of research interest. There are no predetermined lists of questions. The respondent is permitted to take control of the conversations and may digress into areas that he/she considers relevant. The interviewer must, however, possess substantial interviewer skills in order not to allow the respondent to deviate too much from the central issues of the investigation.

The semi-structured and unstructured interviews are recommended where the researcher is undertaking an exploratory study or a study that includes an exploratory element. They allow the researcher the opportunity to "probe" into issues that he may not be immediately aware of and therefore may not be able to capture by available theoretical knowledge.

Interviewing create data quality problems. Common among them are different forms of bias, issues of validity and generalizability. Respondents normally construct and uphold an image or front in relation to the interviewer. As long as they treat the interviewer as a stranger, the information they give will be influenced by this natural caution. Respondents may feel the need to conceal information if they consider its disclosure to be injurious to their career prospects in the organisation. The reality they present through their responses can therefore be a distorted one.

It has also been suggested that interviewers with limited experience can create situations that produce biases in the interview process. These include their body languages and the sequencing of the interview. These biases are particularly very potent in situations of semi and unstructured interviews

Thus, the psychological atmosphere of an interview is at least as important as the mechanics of the interviewing process. When the interviewing context is felt by respondents to be permissive and relaxed, his desire to censor the information he provides will be greatly reduced. Effective interviewing therefore requires insight into the dynamics of interaction. Interviewers are therefore advised to adopt styles that strengthen rapport and goodwill.

Issues of Validity and Reliability in Quantitative Studies

The term *reliability* is used in quantitative research to refer to dependability or consistency in the techniques used (Neuman, 2006). If your research process can be repeated by other students and researchers under identical or similar conditions it will be judged to be reliable. Reliability in this connection is evaluated on three dimensions:

Measurement reliability: This assesses to which extent the variables are measured in a consistent manner.

Stability reliability: This assesses the extent to which the measurement of the variables produces consistent results at different points in time.

Representative reliability: This assesses the extent to which the measurement of the variable yields consistent results for various groups of respondents.

In simple terms, quantitative tools are judged as valid if they measure what they set out to measure. Distinction is usually drawn between two types of validity: construct and content validity. Construct validity entails understanding the theoretical rationale which underlies the measurements derived from specific research. It can be seen as a test of the link between a measure and the underlying theory. Construct validity is usually measured using a correlation coefficient. If a test has construct validity, you would expect to see a reasonable correlation with tests measuring related areas. When the correlation is high, the tool can be considered valid.

Content validity refers to the appropriateness of the research measures used – i.e. whether a tool appears to others to be measuring what it says it does and includes nearly all relevant attributes of the phenomenon under investigation. Face validity is a simple form of content validity. This is done by asking recognized experts in the area to give their opinion on the validity of the tool.

The relationship between reliability and validity can be stated simply as this: For a research measure to be valid, it must also be reliable. Reliability is however a necessary but not sufficient condition for validity.

CHAPTER NINE

MIXED RESEARCH METHODS

Introduction

Your choice of data collection methods must always be guided by your problem formulation. That is, you must use the methods that provide the best opportunities for answering research questions. Some research questions are best answered by using quantitative methods, while others require the use of qualitative methods. An increasing number of researchers, however, consider a combination of quantitative and qualitative methods to provide best insights into several different social science problems. It has been argued that when used along with quantitative methods, qualitative research can help us to interpret and better understand the complex reality of a given situation and the implications of quantitative data. But simply using qualitative and quantitative methods in the same study without thoughtful integration or explanation does not add substantial value to an investigation. Thus, the aim of this chapter is to introduce you to the advantages and disadvantages of mixed methods and how to go about their usage.

First, I summarise the similarities and differences between quantitative and qualitative data methods to enable you to do a quick comparison between the two sets of methods. I then describe the viewpoints that may guide you in your decision to mix the two methods and how to do the combinations.

Similarities and Differences between Quantitative and Qualitative Methods

Table 9.1 provides you with a summary of the main differentiating characteristics of quantitative and qualitative research instruments. It shows that the perspective you adopt in a particular research will determine whether your investigation must be done with quantitative or qualitative methods.

Let us take the example of a study in which you are interested in investigating the use of information technology in an organization. You may choose to emphasise the process of adoption of the information technology and the meanings that employees associate with the technology. Such a focus will suggest the use of qualitative method since this will help you gain insight into the symbolic meanings that workers assign to the information technology through their daily interactions and communication. In this case you will be adopting a constructionist and interpretivist paradigm to your study (see Bryman and Bell, 2011). On the other hand, if your research concerns the types and amounts of data that a company collects from its customers by using various types of computer software, you may be inclined to adopt a quantitative method.

Table 9.1: Some General Characteristics of Quantitative and Qualitative Research

Quantitative Research	Qualitative Research
Test hypothesis that the researcher begins with	Capture and discover meaning once the researcher obtains the data
Concepts are in the form of distinct variables	Concepts are in the form of themes, motifs and taxonomies
Measures are systematically created before data collection and are standardized	Measures are created in an ad hoc manner and often specific to the individual setting or researcher
Data are in the form of numbers from precise measurements	Data are in the form of words and images from documents, observations, and transcripts
Theory is largely causal and deductive	Theory can be causal or non-causal and is often inductive
Procedures are standard, and replication is frequent	Research procedures are particular, and replication is very rare
Analysis proceeds by using statistics, tables or charts and relating them to the hypotheses	Analysis proceeds by extracting themes or generalizations from evidence and organizing data to present a coherent and consistent picture.

Source: Based on Silverman, 1993; Neuman, W.L. (2006) and Bryman and Bell (2011)

Mixing Quantitative and Qualitative Methods

Mixed methods research is generally considered to be a third set of data collection methods. They are therefore described separately from either quantitative or qualitative methods and require careful integration. Truscott *et al* (2010) argue that the goal of mixed methods is not to replace quantitative or qualitative approaches, but to draw from their strengths and minimize their limitations. Similarly, O'Cathain *et al* (2007) argue that mixed methods research is more than mixing different methods. It is a purposeful and powerful blend intended to increase the *yield* of empirical research.

The underlying rationale for considering a mixed method is that individual methods are flawed in one way or the other. But the flaws in each are not identical. This allows researchers to combine methods, not only to gain their individual strengths but also to compensate for their particular faults and limitations. Thus, by approaching a research problem with two or more methods that have non-overlapping weaknesses, one is able to reinforce the quality of the overall knowledge produced by the investigation. The research results are therefore accepted with greater degree of confidence. If the results produce evidence this will suggest to the researcher that the problem under investigation requires further analysis and one must be cautious in interpreting the significance of any one set of data.

Following Green *et al.* (1989) mixed methods will enhance your research in five major ways:

1. *Triangulation* - helps to test the consistency of your findings by using different methods.

2. *Complementarity* – allows you to clarify and illustrate your results by using different methods and providing different perspectives on the issues you have investigated.
3. *Development* – using different methods will enable you to incrementally build on the results obtained from one method by the use of subsequent methods or steps in the research process.
4. *Initiation* – different methods will help bring up new research questions or challenges.
5. *Expansion* – you will obtain greater richness and details in your study by exploring specific features of each method.

Thus, by using mixed methods, you will be able to use pictures and rich narratives to add meaning to statistical information in your study. Alternatively you can provide numbers to add precision to your narratives and pictures. In other words, the strengths of one method can compensate for the weaknesses of another method.

The following steps should guide you in your decision and use of mixed methods.

1. You must convince yourself that a mixed method is appropriate for your research before you decide to use it. This means that you must provide a justification for using mixed methods in the methodology chapter of your thesis.
2. You must carefully and deliberately decide on the types of methods to use and the sequence in which they will be used in your thesis. For example, if you decide to observe the people that you study, by interviewing them and sending them questionnaires to fill, you must decide

on the order in which the different data collection methods will be used and provide reasons for the sequence that you have chosen.

3. Collect the data using the different methods, noting the difficulties you face in the data collection process.
4. Validate the data using the various approaches outlined in chapters 7 and 8
5. Use the data to write your report and reflect on the limitations to your findings (if any) resulting from the data collection methods that you have used.

Triangulation

Triangulation is another common term used in the literature to refer to mixed methods. As Jick (1979) observes, triangulation allows the researcher to improve the accuracy of his conclusions by relying on data from multiple methods. But triangulation means other things than mixing methods. Denzin (1978) identifies four types of triangulation:

- Theoretical triangulation
- Data triangulation
- Investigator triangulation
- Methodological triangulation

Theoretical triangulation is similar to the multiple paradigm approach discussed earlier. In Abnor and Bjerke's terminology, theoretical triangulation allows the researcher to construct his *operative paradigm* in a way that accommodates two or more meta-theoretical perspectives. These perspectives will then guide his data collection and analysis. Harsard's (1991) study noted above is an example of such a study. Students are very often encouraged to engage in theoretical triangulations by combining theories that (together) provide a more useful understanding of the phenomenon they choose to investigate.

Data triangulation is a data collection strategy that derives data from multiple sources or samples. A business student who collects data from different segments of consumers and channel members is invariably engaged in data triangulation. In this regard many eclectic models that guide researchers to seek information about different dimensions of a social phenomenon may be said to motivate the adoption of data triangulation.

Investigator triangulation refers to the use of more than one investigator (i.e. observer, interviewer) or coder and data analyst in a single study. The collected data is considered more credible when those collecting the data do not have prior discussions with each other about the subject of investigation or have in any other way collaborated with each other during the data collection process.

But the most common use of the term triangulation in the literature is what Denzin (1978) refers to as ***methodological triangulation***. It has two forms: (1) within method triangulation, and (2) between method triangulation. When the researcher uses different techniques within the same method it is a *within-method* triangulation. When different methods are employed in the same research (e.g. quantitative and qualitative methods) it is labelled *between-method* triangulation.

Some researchers use the term ***multiple triangulation*** to describe research strategies that use two or more of the categories of triangulation listed above.

Triangulation as a research methodology has been criticised by several social science researchers. Deetz (1996), for example, argues that to assume that different research methods simply provide additive insights into the same phenomenon is an illusion. The thrust of his argument is that the modes of analysis do not work from different points of view on the same thing; they are producing and elaborating in the act of researching different phenomena for different reasons.

An example of Mixed Methods

One of the most frequently cited studies that have used multiple approaches is that of Hassard (1991) who studied work behaviour in a division of the British Fire Service. His research strategy consisted of four different approaches described by Burrell and Morgan (1979) – i.e. *functionalist, interpretive, radical humanist* and *radical structuralist* approaches (see chapter 7). Since each of the four approaches considers some research problems to be more important than others, Hassard modified the focus of his research questions to correspond to the concerns of the various approaches at various stages of the research. A functionalist approach inspired the first part of the study which focused on how firemen assessed the motivating potential of their jobs. To do so he collected quantitative data using the *Job Diagnostic Survey* instrument developed by Hackman and Oldham (1980). The second part of the study relied on an interpretive paradigm in order to gain insight into how routine events in the Fire Service were accomplished in a context of uncertainty which stemmed from the constant threat of emergency calls. He collected the data for this part of the study by asking the firemen to describe and explain their daily tasks in their own words. The third part of the study adopted a radical humanist paradigm. The objective of this part of the study was to understand how management training in the Fire Service contributed towards the reproduction of an ideology that supported and reinforced capitalist values. The fourth part was based on a radical structuralist paradigm and focused on the development of employment relations and conflicts over working time. In this way, he was able to use four different approaches in a single study.

CHAPTER TEN

SUMMARY AND HIGHLIGHTS

In spite of the significant increase in the number of universities and institutions of higher education in Sub-Sahara Africa (SSA), knowledge and competence development remains a major challenge on the sub-continent. I have argued in this book that the reasons for this weak research-based knowledge and competence development are complex. They can be partly attributed to the fact that SSA educational systems do not encourage and equip African students to be curious about nature and society in general and to develop an interest in the pursuit of knowledge and ideas. There are also infrastructural and instructional challenges. For example, some observers argue that inadequately equipped libraries, with limited access to modern journals and the Internet, weak language skills, unavailability of guidelines for students and supervisors, as well as poorly developed research culture are among the challenges that African students face. These observations have motivated me to write this book. This final chapter of the book provides summaries of the arguments in the previous chapters and highlights the key points that you need to bear in mind in your work process.

Problem Formulation or Research Questions

The first task you must engage in when writing a project is to define the research questions you want to address. This is also referred to as *problem formulation*. The problem formulation defines the focus of the project and guides the overall research design. It also shapes the expectations of the readers and examiners of your project. You must therefore give adequate attention and time to this task.

Clarity and precision are essential requirements for good problem formulation. There is a tendency for some students to attempt to formulate their research questions too broadly for fear of not covering the core issues in their subjects. Others tend to have multiple objectives for their projects. You must avoid falling into such traps. You can do so by taking a look at some of the best projects written by past students of your programme and discussing your problem formulation with your supervisor and class mates. These issues have been discussed in details in chapters two and three of the book.

Project Structure and Style of Writing

Chapters four and five highlight the importance of thesis structure and a comprehensive research strategy. A well-structured thesis helps readers follow the arguments that students make and identify the common thread that links the arguments and the logical progression of thoughts underlining them. A good structure also helps you remain focused throughout the writing process and therefore helps to save time. Always make sure that what you write contributes to addressing the research issues your thesis seeks to investigate.You must therefore think through the general sequence and flow of your thoughts when designing your project. You must pay attention to the following:

1. Spend sufficient time on planning the structure of the thesis.

2. Be critical about what you write. The number of pages you write is not the main criterion for assessing the quality of your thesis.

Furthermore, your writing style – i.e. the manner in which you express your thoughts – is also important. It will enhance the clarity of your arguments and the overall quality of your project. You must always remember that the goal of writing the project is to communicate to your readers. To do so effectively, you must write simple and short sentences. You must also avoid borrowing vocabularies and terminologies that you do not understand simply to make your work appear academic.

In terms of strategy I argue that the results you obtain from your investigations will depend on your views on the following four issues:

1. How reality must be perceived – i.e. objective or subjective or a combination of both
2. How you should go about acquiring knowledge about that reality
3. Whether you are interested in finding universal "truths" in your study or you seek to gain unique understanding and interpretations of the social reality you investigate
4. The specific methods and techniques you adopt to gain that knowledge

You must therefore make sure that all the major components of your research design 'fit' with each other. For example, your philosophical preferences must 'fit' with your research goals and purposes as well as your theoretical perspectives. Similarly, your data collection strategy must 'fit' with your research purposes, your conceptual and theoretical framework and your approach to data analysis.

Literature Review and Theories

A critical review of the existing literature in your subject area is essential for you to write a good project. The literature review offers you the opportunity to demonstrate your knowledge of theories, models, and evidence found in the literature and to select those you find particularly useful to your own study. The review therefore enables you to ground your research on available contemporary knowledge. A good review will help you do the following things:

1. You will identify theories and models that are currently used by other researchers in the field.
2. You will be able to summarise what is known and what is not known with respect to the issues that you are interested in – and thereby help you justify your own research.
3. You will be able to identify areas of controversy in the literature and present your views on these controversies
4. You will be able to discuss the strengths and weaknesses that you find in the theories, with particular reference to your own research questions.

I have also drawn your attention to different types of theories (meta-theories, grand theories, mid-range theories and micro theories) and how they can help you sharpen your understanding of the phenomenon that you investigate.

Methods

The discussions in chapters seven to nine show that the methods used in research unavoidably influence the objects of inquiry. For this reason you are advised to provide a clear account of the process of data collection and analysis in the methodology chapter of your project. Although such detailed accounts add to the length of the report, they also allow the readers of your projects to judge whether the findings reported are adequately supported by the data. This will further reinforce the validity and quality of your study.

Chapter seven presents the general characteristics of qualitative research methods. If you define your research task to be one of uncovering meanings rather than testing pre-established hypotheses you must strongly consider adopting qualitative data collection methods. This method allows you to focus on participants' own understanding and interpretations of their situations.

The three qualitative data collection methods that you may consider using are the following:

1. A focus group
2. Observational methods
3. Qualitative interviewing (especially CIT)

A focus group method can be used as a primary source of data or in combination of two or more data gathering methods. Observational methods are useful when the subject of investigation is sensitive and people are reluctant to talk about it or when the issues under investigation can be observed through people's behaviours. The critical incident technique is a useful source of data when the participants in the investigation can identify the events or circumstances that led to the critical incident and when the factors that make the event critical can be identified.

You must NOT choose qualitative methods and techniques because you think that they are simpler to use than quantitative methods. Qualitative data collection and analysis are more time and resource consuming than most quantitative methods. You must choose them when they are consistent with your view of the social phenomenon that you want to investigate (i.e. your problem formulation and paradigmatic preferences).

In chapter eight I argued that quantitative data collection methods and techniques are generally used by researchers that subscribe to an objectivist view of reality and positivist epistemology. This allows them to formulate and test hypotheses and to arrive at results that are generalizable. Thus, if your thesis seeks to produce generalizable knowledge, you must seriously consider using quantitative data collection methods.

There are specific steps and standard procedures that you must follow when collecting quantitative data and some of them have been presented above. There are also various statistical software packages that you can use in analysing quantitative data. You must seek the advice of your supervisor on which packages are available at your university.

Remember that quantitative and qualitative research methods differ primarily in:

- Their analytical objectives
- The types of questions they pose
- The types of data collection instruments they use
- The forms of data they produce
- The degree of flexibility built into study design

Chapter nine discussed the usefulness of mixed methods as a third method of data collection – quantitative and qualitative methods being the other two sets of methods. Remember that mixed methods do not **mean** simply using qualitative and quantitative methods in the same study. Its usage requires clear justification and thoughtful integration. You must therefore provide good reasons for adopting this approach in your thesis.

Bibliography

Allaire, Yvan and Firsirotu, Mihaela E. (1984) "Theories of Organizational Culture" *Organization Studies* Vol. 5 No. 3 pp. 193-226

Arbnor, Ingeman and Bjerke, Björn (2009) *Methodology for Creating Business Knowledge:* London: Sage Publications

Bierman, E. & Jordaan, M.C.E. 2007. Developing applied research skills in 4th year students using e-learning: a case study. Paper Presented at the WWW Applications Conference held from 5-7 September 2007 at the University of Johannesburg, South Africa.

Blumberg, Boris Cooper, Donald R. and Schindler, Pamela S. (2005) *Business Research Methods* 2nd Edition: Boston, McGrawHill Higher Education

Bryman, Alan and Bell, Emma (2011) *Business Research Methods* (Oxford, Oxford University Press)

Burrell, W. G., and Morgan G. (1979) *Sociological Paradigms and Organizational Analysis:* London: Heinemann

Crotty, M. (1998) *The foundations of social science research*: St. Leonards, New South Wales: Allen and Unwin, 1998.

Deetz, S. (1996) "Describing differences in approaches to organization science: Rethinking Burrell and Morgan and their legacy" *Organization Science*, Vol. 7 No. 2: pp. 191–207.

Denzin, N. K. (1978) *The Research Act:* New York: Wiley

Denzin, Norman K. and Lincoln, Yvonna S. (2003) *The Landscape of Qualitative Research* 2nd Edition Thousand Oaks CA Sage Publications,

Deutsch, M., & Krauss, R. M. (1965) *Theories in Social Psychology* (New York: Basic Books).

Dowse, C. & Howie, S. (2013). Promoting academic research writing with South African masters students in the field of education. In Plomp. T & Nieveen, N. (Eds). Educational

Design Research: Introduction and Illustrative Cases. SLO, Netherlands Institute for Curriculum Development, Enschede, The Netherlands

Fisher, Colin (2010) *Researching and Writing a Dissertation* Essex, FT Prentice Hall

Flanagan, J.C., (1954) "The Critical Incident Technique", *Psychological Bulletin* Vol. 51 No.4 pp: 327–58

Greene, Jennifer C., Caracelli, Valerie J. and Graham, Wendy F. (1989) "Toward a conceptual framework for mixed-method evaluation design" *Educational Evaluation and Policy Analysis*, Vol.11 No. 3 pp. 255-274

Gullestrup, Hans (2006) *Cultural Analysis – Towards Cross-cultural Understanding* (Copenhagen: Copenhagen Business School Press)

Hackman, J. R., and Oldham, G. R. (1980) *Work Redesign:* Reading, MA: Addison-Wesley

Hassard, John (1991) "Multiple Paradigms and Organizational Analysis: A Case Study" *Organization Studies* Vol. 12 No. 2 pp: 275-299

Hofstede, G. (2001) *Culture's consequences: Comparing values, behaviors, institutions, and organizations across nations* (2nd Ed.) (Thousand Oaks, California: Sage Publications)

Jick, T.D., (1979) "Mixing qualitative and quantitative methods: Triangulation in action" *Administrative Science Quarterly*, Vol. 24, No. 4, pp: 602-611

Knobel, M. and Lankshear, C. (1999) *Ways of Knowing: Researching Literacy* (Newton: Primary English Teaching Association)

Kuada, John (1994) *Managerial Behaviour in Ghana and Kenya – A Cultural Perspective*, Aalborg, Aalborg University Press

Kuada, John (2012) *Research Methodology – A Project Guide for University Students* (Frederiksberg, Samfundslitteratur)

Kuada, John (2015) *Paradigms and Philosophy of Science – A Doctoral Students' Guide* (London, Adonis & Abbey-Skylark)

Kuada, John and Sørensen, Olav Jull (2000) *Internationalization of Companies from Developing Countries:* New York: Haworth International Business Press

Lovitts, B. (2005) "How to grade a dissertation" *Academe,* Nov/Dec 2005, p. 18-23

Merton, R. K. (1968) *Social Theory and Social Structure* (New York: Free Press)

Morgan, Gareth and Smircich, Linda (1980) "The Case for Qualitative Research" *Academy of Management Review* Vol. 5 No. 4 pp 491-5000

Neuman, Lawrence W. (2006) *Social Research Methods – Qualitative and Quantitative Approaches* 6th Edition: Boston, Pearson Education

O'Cathain, A., Murphy, E., and Nicholl, J. (2007) "Integration and publications as indicators of 'yield' from mixed methods studies" *Journal of Mixed Methods Research* Vol. 1 pp. 147–163.

Rossman, G.B. and Wilson, B.L.(1985) "Numbers and Words: Combining Qualitative and Quantitative Methods in a Single Large Scale Evaluation" *Evaluation Review*, Vol. 9 No.5 pp: 627-643

Saunders, Mark, Lewis, Philip and Thornhill, Andrian (2007) *Research Methods for Business Students* 4th Edition: London, FT Prentice Hall

Silverman, David (2010) *Doing Qualitative Research* (London, Sage Publications)

Strauss, Anselm and Corbin, Juliet (1998) *Basics of Qualitative Research* (London: Sage Publications)

Szanton DL, Manyika S: (2002) *PHD programs in African universities: Current status and future prospects.* Berkeley,

California: The Institute of International Studies and Center for African Studies University of California

Truscott, Diane M. Swars, S., Smith, S., Thornton-Reid, F., Zhao, Y., Dooley, C., Williams, B., Hart. L., and Matthews, M (2010) "A cross-disciplinary examination of the prevalence of mixed methods in educational research: 1995–2005" *International Journal of Social Research Methodology* Vol. 13, No. 4, pp, 317–328

Whetten, David A., (1989) "What constitutes a theoretical contribution?" *Academy of Management Review* Vol. 14 No. 4 pp: 490-495

Index

www.ingramcontent.com/pod-product-compliance
Lightning Source LLC
Chambersburg PA
CBHW070929270326
41927CB00011B/2784